LTE-ADVANCED

LTE-ADVANCED
3GPP SOLUTION FOR
IMT-ADVANCED

Editors

Harri Holma and Antti Toskala
Nokia Siemens Networks, Finland

WILEY

A John Wiley & Sons, Ltd., Publication

This edition first published 2012
© 2012 John Wiley & Sons, Ltd

Registered office
John Wiley & Sons Ltd, The Atrium, Southern Gate, Chichester, West Sussex, PO19 8SQ, United Kingdom

For details of our global editorial offices, for customer services and for information about how to apply for permission to reuse the copyright material in this book please see our website at www.wiley.com.

Library of Congress Cataloguing-in-Publication Data

LTE-advanced : 3GPP solution for IMT-advanced / edited by Harri Holma, Antti Toskala.
    p. cm.
  Includes bibliographical references and index.
  ISBN 978-1-119-97405-5 (cloth)
  1. Long-Term Evolution (Telecommunications)   I. Holma, Harri, 1970–
II. Toskala, Antti.
  TK5103.48325.L73 2012
  621.3845′6–dc23

                                          2012012173

A catalogue record for this book is available from the British Library.

ISBN: 9781119974055

Set in 10/12pt Times by Thomson Digital, Noida, India.

To Kiira and Eevi

– Harri Holma

To Lotta-Maria, Maija-Kerttu and Olli-Ville

– Antti Toskala

Contents

List of Contributors

All contributors from Nokia Siemens Networks unless otherwise indicated.

Mieszko Chmiel
Amitava Ghosh
Kari Hooli
Pasi Kinnunen
Troels Kolding
Jari Lindholm
Timo Lunttila
Patrick Marsch
Klaus Pedersen
Bernhard Raaf
Karri Ranta-aho
Rapeepat Ratasuk
Simone Redana
Claudio Rosa
Cinzia Sartori
Peter Skov
Jun Tan
Hua Wang[*]
Xiaoyi Wang
YuYu Yan

[*] This contributor is from Aalborg University, Denmark.

Preface

The data usage growth in the mobile networks has been very fast during the last few years: the networks have turned rapidly from voice-dominated into data-dominated. The data growth has been fuelled by the availability of mobile broadband coverage and by the higher data rate capabilities. LTE networks were launched early 2009 pushing the data rates up to 100 Mbps. The LTE-capable devices including smartphones and tablet computers became widely available during 2012 boosting the demand for LTE networks and data rates. The further evolution continues on top of LTE, called LTE-Advanced, pushing the data rates beyond 1 Gbps and increasing the system capacity. This book presents 3GPP LTE-Advanced technology in Release 10 and evolution to Release 11 and beyond. The expected practical performance is also illustrated in this book.

The book is structured as follows. Chapter 1 presents an introduction. The standardization schedule and process is described in Chapter 2. An overview of LTE in Release 8 and 9 is

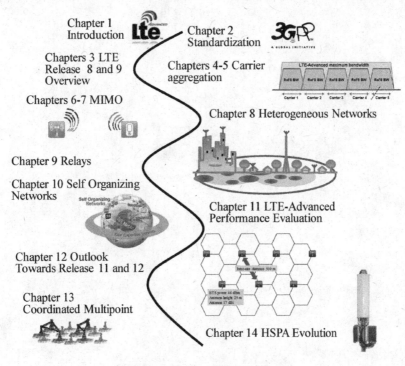

Figure P.1 Contents of the book.

given in Chapter 3. Chapters 4 and 5 present the carrier aggregation solution in downlink and in uplink. Chapters 6 and 7 illustrate the multiantenna Multiple Input Multiple Output (MIMO) techniques in downlink and in uplink. The multilayer and multitechnology hetero-geneous networks are covered in Chapter 8. Chapter 9 introduces relays and their benefits, and Chapter 10 describes Self-Organizing Network (SON) algorithms. The radio perform-ance evaluation is discussed in Chapter 11. The outlook towards future standardization is presented in Chapter 12. The coordinated multipoint concept is illustrated in Chapter 13. Chapter 14 summarizes the latest enhancements in High Speed Packet Access (HSPA) evolution.

Acknowledgements

The editors would like to acknowledge the hard work of the contributors from Nokia Siemens Networks: Mieszko Chmiel, Amitava Ghosh, Kari Hooli, Pasi Kinnunen, Troels Kolding, Jari Lindholm, Timo Lunttila, Patrick Marsch, Klaus Pedersen, Bernhard Raaf, Rapeepat Ratasuk, Karri Ranta-aho, Simone Redana, Claudio Rosa, Cinzia Sartori, Peter Skov, Jun Tan, Hua Wang, Xiaoyi Wang and Yuyu Yan.

We also would like to thank the following colleagues for their valuable comments: Ömer Bulakci, Lars Dalsgaard, Matthias Hesse, Krzysztof Kordybach, Peter Merz, Sari Nielsen, Sabine Rössel and Hanns Jürgen Schwarzbauer.

The editors appreciate the fast and smooth editing process provided by the publisher, John Wiley & Sons, Ltd and especially Mariam Cheok, Richard Davies, Sandra Grayson and Mark Hammond.

We are grateful to our families, as well as the families of all the authors, for their patience during the late night and weekend editing sessions.

The editors and authors welcome any comments and suggestions for improvements or changes that could be implemented in forthcoming editions of this book. The feedback is welcome to editors' e-mail addresses harri.holma@nsn.com and antti.toskala@nsn.com.

List of Abbreviations

3GPP	Third Generation Partnership Project
AAA	Authentication, Authorization and Accounting
ABS	Almost Blank Subframe
ACK	Acknowledgement
ACLR	Adjacent Channel Leakage Ratio
ADC	Analogue-to-Digital Conversion
ADSL	Asymmetric Subscriber Line
A-GW	Access Gateway
AM	Acknowledged Mode
AMC	Adaptive Modulation and Coding
AMR	Adaptive MultiRate
AMR-NB	AMR NarrowBand
AMR-WB	AMR WideBand
ANDSF	Access Network Discovery and Selection Function
AP	Application Protocol
ASN.1	Abstract Syntax Notation One
CA	Carrier Aggregation
CAPEX	Capital Expenditures
CB	Coordinated Beamforming
CC	Component Carrier
CCO	Coverage and Capacity Optimization
CCS	Component Carrier Selection
CDD	Cyclic Delay Diversity
CDM	Code Division Multiplex
CDMA	Code Division Multiple Access
CGI	Cell Global Identity
CIF	Carrier Indicator Field
CM	Cubic Metric
CO2	Carbon Dioxide
COC	Cell Outage Compensation
CoMP	Coordinated MultiPoint
CPC	Continuous Packet Connectivity
CQI	Channel Quality Indicator
CRS	Common Reference Signals

CS	Coordinated Scheduling
CS	Circuit Switched
CSG	Closed Subscriber Group
CSI	Channel State Information
CSoHSPA	Circuit Switched Voice over HSPA
DAC	Digital-to-Analogue Conversion
DAS	Distributed Antenna System
DCC	Downlink Component Carrier
DCCH	Dedicated Control Channel
DCH	Dedicated Channel
DC-HSDPA	Dual Cell HSDPA
DCI	Downlink Control Information
DCS	Dynamic Cell Selection
DIP	Dominant Interferer Proportion
DM-RS	Demodulation Reference Signal
DPCCH	Dedicated Physical Control Channel
DRX	Discontinuous Reception
DTX	Discontinuous Transmission
DwPTS	Downlink Pilot Time Slot
eICIC	Enhanced Inter-Cell Interference Coordination
EIRP	Equivalent Isotropic Radiated Power
eNB	eNodeB
EPC	Evolved Packet Core
ePDCCH	enhanced Physical Downlink Control Channel
ES	Energy Saving
FACH	Forward Access Channel
FDD	Frequency Division Duplex
FGI	Feature Group Indicators
GGSN	Gateway GPRS Support Node
GNSS	Global Navigation Satellite System
GP	Guard Period
GPRS	General Packet Radio Service
GSM	Global System for Mobile Communication
GTP	GPRS Tunnelling Protocol
GW	Gateway
HARQ	Hybrid Automatic Repeat-reQuest
HetNet	Heterogeneous Networks
HLR	Home Location Register
HO	Handover
HSDPA	High Speed Downlink Packet Access
HS-FACH	High Speed FACH
HSPA	High Speed Packet Access
HSS	Home Subscriber Server
HSUPA	High Speed Uplink Packet Access
IC	Interference Cancellation
ICIC	Inter-Cell Interference Coordination

ID	Identity
IMS	IP Multimedia Sub-system
IMT	International Mobile Telecommunications
IP	Internet Protocol
IPSec	IP Security
IQ	Imaginary Quadratic
IRC	Interference Rejection Combining
ISD	Inter-Site Distance
ITU	International Telecommunication Union
JT	Joint Transmission
KPI	Key Performance Indicator
LGW	Local Gateway
LIPA	Local IP Access
LLR	Log-likelihood Ratio
LTE	Long Term Evolution
MAC	Medium Access Control
MBMS	Multimedia Broadcast Multicast Service
MBSFN	Multicast Broadcast Single Frequency Network
MCL	Minimum Coupling Loss
MDT	Minimization of Drive Testing
MIB	Master Information Block
MIMO	Multiple Input Multiple Output
MLB	Mobility Load Balancing
MME	Mobility Management Entity
MRC	Maximal Ratio Combining
MRO	Mobility Robustness
MMSE	Minimum Mean Square Error
MPR	Maximum Power Reduction
MTC	Machine Type Communication
MU-MIMO	Multi-User MIMO
OAM	Operation Administration Maintenance
OCC	Orthogonal Cover Codes
OFDM	Orthogonal Frequency Division Multiplexing
O&M	Operation and Maintenance
OPEX	Operating Expenditures
OTDOA	Observed Time Difference Of Arrival
PA	Power Amplifier
PBCH	Physical Broadcast Channel
PCC	Primary Component Carrier
PCC	Policy and Charging Control
PCell	Primary Cell
PCFICH	Physical Control Format Indicator Channel
PCH	Paging Channel
PCI	Physical Cell Identity
PCRF	Policy and Charging Resource Function
PDCCH	Physical Downlink Control Channel

PDCP	Packet Data Convergence Protocol
PDN	Packet Data Network
PDP	Packet Data Protocol
PDU	Payload Data Unit
P-GW	PDN Gateway
PHICH	Physical HARQ Indicator Channel
PHR	Power Headroom Reporting
PIC	Parallel Interference Cancellation
PMI	Precoding Matrix Indicator
PRB	Physical Resource Block
PRG	Precoding Resource block Group
PSS	Primary Syncronization Signal
PUCCH	Physical Uplink Control Channel
PUSCH	Physical Uplink Shared Channel
QAM	Quadrature Amplitude Modulation
QCI	QoS Class Identifier
QoS	Quality of Service
QPSK	Quadrature Phase Shift Keying
RACH	Random Access Channel
RAN	Radio Access Network
RAT	Radio Access Technology
RE	Range Extension
RER	Re-Establishment Request
RET	Remote Electrical Tilt
RF	Radio Frequency
RI	Rank Indicator
RIM	RAN Information Management
RLC	Radio Link Control
RLF	Radio Link Failure
RN	Relay Node
RNC	Radio Network Controller
RNTP	Radio Network Temporary Identifier
R-PDCCH	Relay PDCCH
RRC	Radio Resource Control
RSRP	Reference Signal Received Power
RSRQ	Reference Signal Received Quality
RTP	Real Time Protocol
RRH	Remote Radio Head
RRM	Radio Resource Management
SA	System Aspects
SAE	System Architecture Evolution
SAP	Single Antenna Port
SCell	Secondary Cell
SFN	Single Frequency Network
SGSN	Serving GPRS Support Node
S-GW	Serving Gateway

SIB	System Information Block
SINR	Signal to Interference and Noise Ratio
SON	Self-Organizing Networks
SORTD	Space-Orthogonal Resource Transmit Diversity
SPS	Semi-Persistent Scheduling
SRS	Sounding Reference Signal
SR-VCC	Single Radio Voice Call Continuity
SSS	Secondary Synchronization Signal
TCO	Total Cost of Ownership
TDD	Time Division Duplex
TM	Transmission Mode
TTG	Tunnel Termination Gateway
TTI	Transmission Time Interval
UCI	Uplink Control Information
UDP	User Datagram Protocol
UE	User Equipment
ULA	Uniform Linear Arrays
UM	Unacknowledged Mode
URS	UE specific Reference Signal
UpPTS	Uplink Pilot Time Slot
USB	Universal Serial Bus
VoIP	Voice over IP
VoLTE	Voice over LTE
WCDMA	Wideband Code Division Multiple Access
WiFi	Wireless Fidelity
WiMAX	Worldwide Interoperability for Microwave Access
WLAN	Wireless Local Area Network

1

Introduction

Harri Holma and Antti Toskala

1.1 Introduction

The huge popularity of smartphones and tablet computers has pushed the need for mobile broadband networks. Users find increasing value in mobile devices combined with a wireless broadband connection. Users and new applications need faster access speeds and lower latency while operators need more capacity and higher efficiency. LTE is all about fulfilling these requirements. GSM made voice go wireless, HSPA made initial set of data connections go wireless and now LTE offers massive capabilities for the mobile broadband applications.

The first set of LTE specifications were completed in 3GPP in March 2009. The first commercial LTE network opened in December 2009. There were approximately 50 commercial LTE networks by the end of 2011 and over 100 networks are expected by the end of 2012. The first LTE smartphones were introduced in 2011 and a wide selection of devices hit the market during 2012. An example LTE smartphone is shown in Figure 1.1: the Nokia 900 with 100 Mbps LTE data rate and advanced multimedia capabilities. Overall, LTE technology deployment has been a success story. LTE shows attractive performance in the field in terms of data rates and latency and the technology acceptance has been very fast. The underlying technology capabilities evolve further which allows pushing also LTE technology to even higher data rates, higher base station densities and higher efficiencies. This book describes the next step in LTE evolution, called LTE-Advanced, which is set to increase the data rate even beyond 1 Gbps.

1.2 Radio Technology Convergence Towards LTE

The history of mobile communications has seen many competing radio standards for voice and for data. LTE changes the landscape because all the existing radios converge towards LTE. LTE is the evolution of not only GSM/HSPA operators but also CDMA and WiMAX operators. Therefore, LTE can achieve the largest possible ecosystem. LTE co-exists smoothly with the current radio networks. Most GSM/HSPA operators keep their existing

LTE-Advanced: 3GPP Solution for IMT-Advanced, First Edition. Edited by Harri Holma and Antti Toskala.
© 2012 John Wiley & Sons, Ltd. Published 2012 by John Wiley & Sons, Ltd.

Figure 1.1 An example of an LTE smartphone – Nokia Lumia 900.

GSM and HSPA radio networks running for long time together with LTE, and they also keep enhancing the existing networks with GSM and HSPA evolutions. The LTE terminals are multimode capable supporting also GSM and HSPA. The radio network solution is based on multi-radio base station which is able to run simultaneously all three radios. Many operators introduce multi-radio products to their networks together with LTE rollouts to simplify the network management and to modernize the existing networks.

The starting point for CDMA and WiMAX operators is different since there is no real evolution for those radio technologies happening. Therefore, CDMA and WiMAX operators tend to have the most aggressive plans for LTE rollouts to get quickly to the main stream 3GPP radio technology to enjoy the LTE radio performance and to get access to the world market terminals.

The high level technology evolution is illustrated in Figure 1.2.

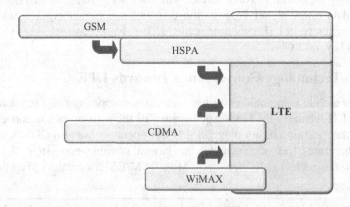

Figure 1.2 Radio technology convergence towards LTE.

1.3 LTE Capabilities

LTE Release 8 offers peak data rate of 150 Mbps in downlink by using 20 MHz of bandwidth and 2×2 MIMO. The first LTE devices support up to 100 Mbps while the network capability is up to 150 Mbps. The average data rates in the commercial networks range between 20 and 40 Mbps in downlink and 10–20 Mbps in uplink with 20 MHz bandwidth. Example drive test results are shown in Figure 1.3. Practical LTE data rates in many cases are higher than the available data rates in fixed Asymmetric Digital Subscriber Lines (ADSL). LTE has been deployed using number of different bandwidths: most networks use bandwidth from 5 to 20 MHz. If the LTE bandwidth is smaller than 20 MHz, the data rates scale down correspondingly. LTE has been rolled out both with Frequency Division Duplex (FDD) and Time Division Duplex) TDD variants. LTE has the benefit that both the FDD and TDD modes are highly harmonized in standardization.

The end user performance is also enhanced by low latency: the LTE networks can offer round trip times of 10–20 ms. The LTE connections support full mobility including seamless intra-frequency LTE handovers and inter-RAT (Radio Access Technology) mobility between LTE and legacy radio networks. The terminal power consumption is optimized by using discontinuous reception and transmission (DRX/DTX).

LTE also offers benefits for the operators in terms of simple network deployment. The flat architecture reduces the number of network elements and the interfaces. Self-Organizing Network (SON) has made the network configuration and optimization simpler enabling faster and more efficient network rollout.

LTE supports large number of different frequency bands to cater the needs of all global operators. The large number of RF bands makes it challenging to make universal LTE devices. The practical solution is to have several different device variants for the different markets. The roaming cases are handled mainly by legacy radios.

Initial LTE smartphones have a few different solutions for voice: Circuit Switched Fallback (CSFB) handover from LTE to legacy radio (GSM, HSPA, CDMA) or dual radio CDMA + LTE radio. Both options use the legacy circuit switched network for voice and

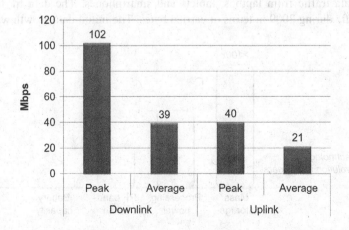

Figure 1.3   Example drive test data rates in LTE network with 20 MHz bandwidth.

LTE network for data. The Voice over LTE (VoLTE) solution with Voice over IP (VoIP) also started during 2012.

1.4 Underlying Technology Evolution

The radio technology improvements need to be supported by the evolution of the underlying technologies. The technology components – including mass storage, baseband, RF and batteries – keep evolving and help the radio improvements to materialize. The size of the mass storage is expected to have fastest growth during the next ten years which allows for storing more data on the device and which may fuel data download over the radio. The memory size can increase from tens of Gigabytes to several Terabytes. Also the digital processing has its strong evolution. The digital processing power has improved according to Moore's law for several decades. The evolution of the integration level will not be as easy as in earlier times, especially when we need to minimize the device power consumption. Still, the digital processing capabilities will improve during the 2010s, which allows for processing of higher data rates and more powerful interference cancellation techniques. Another area of improvement is the RF bandwidth which increases mainly because of innovations in digital front end processing. The terminal power consumption remains one of the challenges because the battery capacity is expected to have relatively slow evolution. Therefore, power saving features in the devices will still be needed. The technology evolution is illustrated in Figure 1.4.

LTE-Advanced devices and base stations will take benefit of the technology evolution. Higher data rates and wider bandwidth require baseband and RF evolution. The attractive LTE-Advanced devices also benefit from larger memory sizes and from improved battery capacity.

1.5 Traffic Growth

The data volumes in mobile networks have increased considerably during the last few years and the growth is expected to continue. The traffic growth since 2007 and the expected growth until 2015 are illustrated in Figure 1.5. The graph shows the total global mobile network data volume in Exabytes; that is, millions of Terabytes. The traffic is split into voice traffic and data traffic from laptops, tablets and smartphones. The data traffic exceeded the voice traffic during 2009 in terms of carried bytes. The initial data growth was driven by

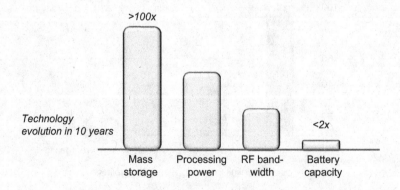

Figure 1.4   Evolution of underlying technology components.

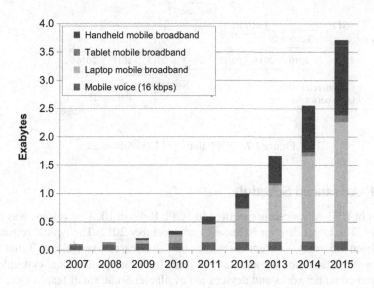

Figure 1.5 Expected traffic growth (Nokia Siemens Network estimate 2011).

the laptop modems; see an example in Figure 1.6. It is also expected that the LTE-Advanced capabilities, like higher data rates, are first introduced for the laptop modems. The relatively fastest growth from 2012 to 2015 is expected to come from smartphones. The smartphones make nearly half of the traffic by 2015. The total traffic by 2015 will be approximately 40 times more than the traffic 2007. The share of voice traffic is expected to shrink to less than 5% by 2015. Some of the advanced markets already have the total traffic 50 times more than the voice traffic; that means voice is less than 2% of total traffic.

It is not only the data volume that is growing in the networks but also the amount of signalling grows and the number of connected devices grows. The radio evolution work needs to address all these growth factors.

Figure 1.6 Example of a 100 Mbps USB modem – Nokia Siemens Networks 7210.

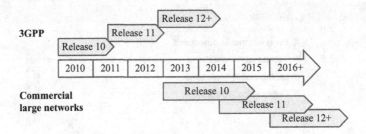

Figure 1.7 3GPP timing of LTE-Advanced.

1.6 LTE-Advanced Schedule

The first set of LTE-Advanced is specified in 3GPP Release 10. That release was completed in June 2011. The target date for Release 11 is December 2012. The typical release cycle in 3GPP has been 1.5 years – except for some smaller releases like Release 9 that was completed in a year. It tends to take another 1.5 years from the specification's completion until the first commercial networks and devices are available. Some small features can be implemented faster while some major features requiring heavy redesign may take more time. We could then expect that the first LTE-Advanced features are commercially available during 2013, and Release 11 features towards end of 2014. The LTE-Advanced schedule is shown in Figure 1.7.

1.7 LTE-Advanced Overview

The main features of LTE-Advanced are summarized in Figure 1.8.

- Downlink carrier aggregation to push the data rate initially to 300 Mbps with $20 + 20$ MHz spectrum and 2×2 MIMO, and later to even 3 Gbps by using 100 MHz bandwidth and 8×8 MIMO. More bandwidth is the handy solution to increase the data rates.
- Multiantenna MIMO evolution to 8×8 in downlink and 4×4 in uplink. The multiantenna MIMO can also be used at the base station while keeping the number of terminal antennas low. This approach offers the beamforming benefits increasing the network capacity while

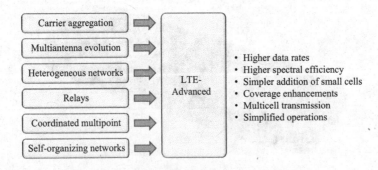

Figure 1.8 Overview of LTE-Advanced main features.

keeping the terminal complexity low. Multiantennas increase the data rates and the network capacity.

- Heterogeneous network (HetNet) for the co-channel deployment of macrocells and small cells. HetNet features enable interference coordination between the cell layers. Those features enhance the network capacity and coverage with high density of small cells while sharing the frequency with large macrocells.
- Relay nodes for backhauling the base stations via LTE radio interface. The transmission link can use inband or outband transmission. Relays are practical for increasing network coverage if the backhaul connections are not available.
- Coordinated multipoint transmission and reception allows using several cells for the data connection towards one terminal. Coordinated multipoint improves especially the cell edge data rates that are limited by inter-cell interference.
- Self-organizing network features make the network rollout faster and simpler, and improves the end user performance by providing correct configurations and optimized parameter setting.

LTE-Advanced features in Release 10 can be upgraded flexibly on top of Release 8 network on the same frequencies while still supporting all legacy Release 8 terminals. Therefore, the evolution from LTE to LTE-Advanced will be a smooth one. All these features will be described in detail in this book.

1.8 Summary

LTE Release 8 has turned out to be a successful technology in terms of practical performance and in terms of commercial network and terminal launches. At the same time the high popularity of smartphones pushes the need for further mobile broadband evolution. LTE-Advanced is designed to enhance LTE capabilities in terms of data rates, capacity, coverage and operational simplicity. The first set of LTE-Advanced specifications was completed in 3GPP during 2011 and the features are expected to be commercially available 2013. LTE-Advanced is backwards compatible with LTE and can co-exist with LTE Release 8 terminals on the same frequency.

2

LTE-Advanced Standardization

Antti Toskala

2.1 Introduction

This chapter presents the Long Term Evolution (LTE)-Advanced standardization aspects. The standardization of LTE-Advanced is handled by the 3rd Generation Partnership Project (3GPP). 3GPP produced also the earlier LTE versions, Release 8 and 9, as well as the Wideband CDMA (WCDMA) and High Speed Packet Access (HSPA) Releases (and later GSM evolution Releases also). 3GPP has become the leading standards forum for the 4th generation mobile communication systems, following the wide adoption of LTE and LTE-Advanced technology path as the future choice by all the major operators from WCDMA/HSPA, CDMA and WiMAX technology camps. The 3GPP consists of all the major vendors and equipment manufacturers, including all the global players providing infrastructure, terminals, chipsets or wireless test equipment. The 3GPP is consisting of 384 member organizations and approximately 4000 delegate days every month invested in progressing the technology with around 9000 change requests approval to the specifications annually. Tens of thousands of input documents are submitted from different companies to different 3GPP working groups each year. The 3GPP structure and process is covered more detail in [1]. This chapter introduces the relevant 3GPP Release schedule for LTE-Advanced and then presents an overview of the LTE-Advanced study phase done before the actual specification work started in 3GPP. The requirements set for LTE-Advanced by the 3GPP community, as well as ITU-R for the IMT-Advanced submission process, are reviewed and then foreseen steps for later LTE Releases are covered. This chapter concludes with the introduction of relevant 3GPP specifications for LTE-Advanced.

2.2 LTE-Advanced and IMT-Advanced

The International Telecommunication Union Radiocommunication Sector (ITU-R) [2] started process for the new system, International Mobile Telecommunications Advanced (IMT-Advanced), with the circular letter distributed in 2008 to call for radio technology

LTE-Advanced: 3GPP Solution for IMT-Advanced, First Edition. Edited by Harri Holma and Antti Toskala.
© 2012 John Wiley & Sons, Ltd. Published 2012 by John Wiley & Sons, Ltd.

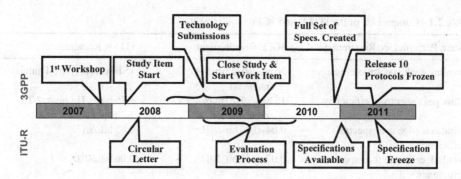

Figure 2.1 LTE-Advanced standardization and IMT-Advanced process schedule.

proposals, following the earlier process for IMT-2000 as covered in [1]. ITU-R called for proposals for radio technologies that could meet the requirements set for a versatile radio technology to qualify as an IMT-Advanced technology, often also denoted as the 4th generation (4G) mobile communication system. 3GPP responded to the circular letter with the submission of the Release 10 LTE-Advanced and evaluation results of the achievable performance with LTE-Advanced. The process was completed with the full set of specifications submitted at the end of 2010, with some updates during 2011. The ITU-R IMT-Advanced and 3GPP LTE-Advanced process schedules are shown in Figure 2.1.

2.3 LTE-Advanced Requirements

The ITU-R defined the requirements for the IMT-Advanced such that the system should be able to [3]:

- Enable 100 Mbps peak data rate support for high mobility and up to 1 Gbps peak data rate for low mobility case;
- Allow inter-working with other radio access systems;
- Enabling high quality mobile services;
- Worldwide roaming capability;
- Flexibility to allows cost efficient support of wide range of service and applications.

Further there were requirements in the following areas:

- Cell spectral efficiency, ranging from the 3 bits/Hz/cell in the indoor downlink scenario, to the 0.7 bits/Hz/cell in the high speed uplink scenario;
- Peak spectral efficiency, ranging up to 15 bits/s/Hz;
- Bandwidth scalability up to and including 40 MHz, up to 100 MHz should also be considered;
- Cell edge user spectral efficiency, ranging from 0.015 bits/s/Hz up to 0.1 bits/s/Hz;
- Latency requirements for control plane to achieve 100 ms transition time between idle and active state, and respectively to enable 10 ms user plane latency (in unloaded conditions);
- Mobility support with up to 350 kmph (smaller data rate allowed compared to the stationary use case);

Table 2.1 Comparison of the ITU-R and 3GPP requirements

System Performance Requirements	3GPP Requirement	ITU-R Requirement
Downlink peak spectrum efficiency	30 bits/s/Hz (max 8 antennas)	15 bits/s/Hz (max 4 antennas)
Uplink peak spectrum efficiency	15 bits/s/Hz (max 4 TX antennas)	6.75 bits/s/Hz (max 2 TX antennas)
Uplink cell edge user spectral efficiency	0.04–0.07 bits/s/Hz	0.03 bits/s/Hz
Downlink cell edge user spectral efficiency	0.07–0.12 bits/s/Hz	0.06 bits/s/Hz
User plane latency	10 ms	10 ms

- Handover interruption of 27.5 ms for intra frequency case and 40 and 60 ms for the inter-frequency within the band and between the bands respectively;
- VoIP capacity, with the numbers of users ranging from 30 to 50 users per sector/MHz.

3GPP also defined its own requirements for LTE-Advanced, which are in many cases tighter than the requirements for IMT-Advanced. This was due to the Release 8 LTE being quite an advanced system already, so that it could meet the IMT-Advanced requirements in many areas already (especially in the high mobility cases). Thus there was desire to ensure in 3GPP that there would be sufficient incremental steps between Release 8/9 LTE capabilities and Release 10 LTE-Advanced capabilities and performance.

Table 2.1 summarizes the key performance requirements from ITU-R and 3GPP [4], indicating the difference especially on the peak spectrum efficiency and on the cell edge spectral efficiency requirements. The latency requirement was identical in both cases. A more detailed treatment on the requirements and achievable performance can be found from Chapter 11.

2.4 LTE-Advanced Study and Specification Phases

The first part of the work in 3GPP was the study phase, which started in early 2008 with the findings on physical layer covered in [5] and overall conclusions presented in [6]. There were several areas studied as part of the study phase, including:

- Achievable capacity and cell edge performance, both for data and VoIP;
- Latency, both for control plane and user plane;
- Handover performance;
- Achievable peak spectral efficiency (and peak data rates);
- Radio Frequency (RF) aspects of supporting the considered technologies.

The results from 3GPP showed that the technology components under consideration can meet or exceed the requirements, as shown with the study phase reports in Chapter 11.

Following the study phase, the work item phase (when actual specifications are produced) was started which produced the first full set of specifications at the end of 2010. This is part of the 3GPP Release 10 specifications, which were finalized in June 2011.

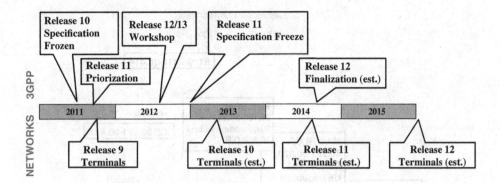

Figure 2.2 LTE-Advanced upcoming release schedule and expected market deployment.

2.5   Further LTE-Advanced 3GPP Releases

3GPP continues the work full speed ahead after the first LTE-Advanced Release. Release 11 work followed immediately Release 10 activity with plenty of proposed work items to further enhance the LTE-Advanced capabilities from Release 10. In order to maintain the schedule 3GPP had to take special actions to prioritize the Release 11 content in September 2011. The selected content is expected to be finalized during second half of 2012 with the specification freeze scheduled to take place at the end of 2012, as shown in Figure 2.2. The key areas where further work was identified included carrier aggregation enhancements and enhanced downlink control channel, as explained with the Release 11 content in more detail in Chapter 12.

3GPP is preparing also the Release 12 and beyond LTE-Advanced content. The 3GPP TSG RAN Release 12/13 workshop in June 2012 collected the vision of the content of the Releases ahead from 3GPP operators and manufacturers. The official Release 12 timing has not yet been decided though with the typical 18-month Release duration and Release 12 could be expected to finalize during the second half of 2014. Figure 2.2 shows also the estimated milestones for the Release 10, 11 and 12 terminals to enter the market based on the assumption of the availability of the first implementations 18 months after the freeze of the protocol specifications (ASN.1 protocol language start of backwards compatibility, as explained in details in [2]) of the corresponding Release.

2.6   LTE-Advanced Specifications

Following the 3GPP Release principle, the LTE-Advanced technology components were added in Release 10 version of the existing LTE specifications, which already contained the Release 8 and 9 LTE features. In some areas new specifications were created, such as for the physical layer of the relay backhaul operation [7] as shown in Figure 2.3, while for example the physical layer impacts of carrier aggregation or multiple antenna enhancements were captured in the existing 36.2xx specification series. The carrier aggregation band combinations are Release independent, with each band combination done as a separate work item, as shown in [8] as an example. The carrier aggregation band combinations can be implemented on top of Release 10 if there is no need for any of the Release 11 specific features. If some of

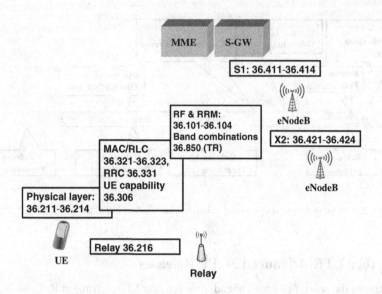

Figure 2.3 LTE-Advanced in 3GPP specifications.

the Release 11 features would be needed by the deployment scenario, such as multiple uplink timing advance values, then the band combination should be based on Release 11 version of the LTE-Advanced specification. With the same principle a new frequency band can be added in 3GPP specifications in a Release independent way so that if a new band is finalized during Release 11 or 12 timeframe one does not need to wait for the ongoing Release to be completed and terminals to be available, but one can implement still for example Release 8 based terminal to the new frequency band (or Release 10 based if LTE-Advanced features are desired) as long as fulfilling the band specific RF and performance requirements.

2.7 Conclusions

In the previous sections, we have covered the LTE-Advanced standardization aspects. The 3GPP has really become the spearhead of development in mobile radio technology with a large attendance from all over the world. The 3GPP is continuing toward further development steps of LTE-Advanced with the ongoing Release 11 work and with the preparation of the plans for Release 12 and 13 as summarized in [9]. Continuous evolution work ensure the LTE-Advanced will stay the most advanced solution for the mobile operators worldwide which is demonstrated by the large participation of the operators in the LTE standardization work from all major markets and with varying technology backgrounds in terms of legacy systems deployed.

References

1. Holma, H. and Toskala, A. (2010) *WCDMA for UMTS*, 5th edn, John Wiley & Sons, Ltd, Chichester.
2. ITU-R Home Page. Available at: http://www.itu.int/ITU-R (accessed May 2, 2012).
3. ITU-R report, M.2134 (November 2008), Requirements related to technical performance for IMT-Advanced radio interface(s).

4. 3GPP technical report TR 36.913 (March 2009) Requirements for further advancements for Evolved Universal Terrestrial Radio Access (E-UTRA) (LTE-Advanced), V8.0.1.
5. 3GPP technical report TR 36.814 (March 2010), Feasibility study for Further Advancements for E-UTRA (LTE-Advanced).
6. 3GPP technical report TR 36.912 (March 2010) Feasibility study for Further Advancements for E-UTRA (LTE-Advanced), v 9.2.0.
7. 3GPP technical specification TS 36.216 (September 2010) Physical layer for relaying operation, V10.0,0.
8. 3GPP Tdoc RP-100668 (June 2010) Work Item Description: LTE-Advanced Carrier Aggregation of Band 3 and Band 7, TeliasSonera.
9. 3GPP Tdoc RWS-120045 (June 2012) Summary of TSG-RAN workshop on Release 12 and onward, TSG-RAN Chairman.

3

LTE Release 8 and 9 Overview

Antti Toskala

3.1 Introduction

This chapter presents the overview of Long Term Evolution (LTE) Release 8 and 9. The principles of the first two LTE Releases produced by 3GPP before LTE-Advanced in Release 10 are presented. In many areas of the LTE-Advanced Release 10, the design is based on the Release 8 and 9 principles with only slight modifications or enhancements to improve the performance. Especially the architecture and protocol solutions in many cases are actually unchanged with LTE-Advanced, perhaps only adding necessary elements to activate the introduced LTE-Advanced physical layer features. This chapter first introduces the LTE physical layer principles and then continues to cover the architecture and protocols solutions in Release 8 and 9. This chapter continues further to cover the overview of the Evolved Packet Core (EPC) and IP Multimedia System (IMS). This chapter is concluded with the UE capability and introduction of the differences between Release 8 and Release 9.

3.2 Physical Layer

The LTE multiple access is based on Orthogonal Frequency Division Multiple Access (OFDMA) in the downlink direction and on the Single Carrier Frequency Division Multiple Access (SC-FDMA) in the uplink direction.

The OFDMA parameterization is based on the 15 kHz sub-carrier spacing to ensure sufficient robustness to large velocity and frequency error, while at the same time the resulting sampling rates allow to have compatibility with the WCDMA sampling rates to facilitate easier multimode implementation both in the UE and the network side. The ODFMA transmitter and receiver chain example is shown in Figure 3.1.

In the uplink direction the SC-FDMA facilitates power efficient terminal transmitter implementation since there are no parallel waveforms transmitted but the transmission

LTE-Advanced: 3GPP Solution for IMT-Advanced, First Edition. Edited by Harri Holma and Antti Toskala.
© 2012 John Wiley & Sons, Ltd. Published 2012 by John Wiley & Sons, Ltd.

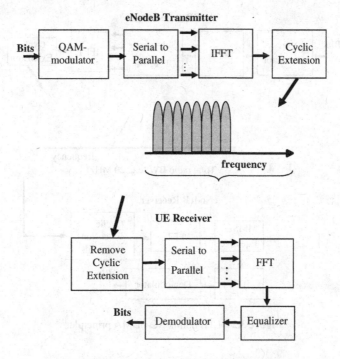

Figure 3.1 LTE downlink OFDMA principle.

principle is based on the use of a digital QAM modulation coupled with the cyclic prefix use after a block of symbols. The Peak to Average Ratio (PAR) or more specifically Cubic Metric (CM) is lower with SC-FDMA than with OFDMA, thus allowing avoiding the use of efficient power amplifier without excessive power back-off to maximize the uplink range, as presented in more details in [1].

Adding a cyclic prefix after every QAM symbol is not feasible due to the short symbol duration in time, which would result in massive overheads. Now the addition of the cyclic prefix is done after the block of symbols which was equal duration to a single OFDMA symbol in the downlink. Thus the uplink and downlink overheads from the use of cyclic prefix are identical. An example of the LTE uplink SC-FDMA transmitter and receiver chain is shown in Figure 3.2. The transmitter uses FFF and IFFT pair to enable the frequency division by allowing placing the transmitted signal easily and efficiently in the instructed part of the uplink bandwidth.

The LTE bandwidths supported in Release 8 and 9 are 1.4, 3, 5, 10, 15 and 20 MHz. In the downlink direction the bandwidth is filled with 15 kHz sub-carriers still leaving room for the necessary reduction of the waveform from the neighbouring carrier. For example with 10 MHz bandwidth there are 600 sub-carriers, thus corresponding to 9 MHz of fully used spectrum. The actual resolution of the resource allocation in frequency domain, both for the uplink and downlink, is 180 kHz. The resource allocation over 180 kHz and for the 1 ms sub-frame corresponds to a single LTE Physical Resource Block (PRB). This is equal to 12 sub-carriers. The parameterization for different bandwidths is shown in Table 3.1. The smallest

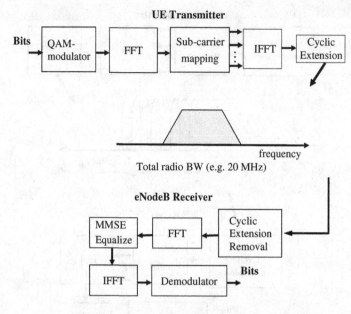

Figure 3.2 LTE uplink SC-FDMA principle.

allocation is thus 6 PRBs, equal to 1.08 MHz and the largest 100 PRBs, equal to 18 MHz. The corresponding bandwidths for normal deployment are then 1.4 and 20 MHz respectively to allow reaching the required values for example for the interference to the adjacent channel, Adjacent Channel Leakage Ratio (ACLR).

The multiple access principle with the use of 180 kHz physical resource block is shown in Figure 3.3 for the downlink direction. An eNodeB is allocating resources every 1 ms for those UEs that have downlink data to be transmitted. The eNodeB scheduler will determine when to transmit data to each UE based on different criteria, including amount of data in the buffer, user priority, service type and momentary channel conditions.

In the uplink direction the operation is similar; the eNodeB will inform each UE whether it can transmit in the uplink direction with the allocation received on the Physical Downlink Control Channel (PDCCH). If UE receives the allocation, it contains information on where

Table 3.1 LTE physical layer bandwidth options and bandwidth specific parameters

Bandwidth	1.4 MHz	3.0 MHz	5 MHz	10 MHz	15 MHz	20 MHz
Sub-frame duration	1 ms					
Sub-carrier spacing	15 kHz					
FFT length	128	256	512	1024	1536	2048
Sub-carriers	72	180	300	600	900	1200
Symbols per slot	7 with Short CP and 6 with Long CP					
Cyclic prefix	5.21 μs with Short CP and 16.67 μs with Long CP					

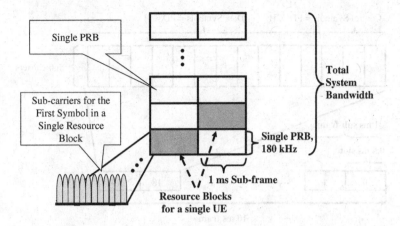

Figure 3.3 Downlink multiple access principle.

in frequency domain the UE is able to transmit the data. In uplink the allocation is always continuous n times 180 kHz, while in the downlink direction and allocation could be with one or two gaps (latter with larger bandwidths only). The uplink operation is shown in Figure 3.4, with the UEs receiving every ms allocation whether they are allowed to transmit or not (and in which part of the frequency) in the following uplink sub-frame.

The frame structure is based on the 10 ms frame which then contains 1 ms sub-frames (that are the equal to the resource allocation period). As shown in Figure 3.5, the sub-frame is divided between control and data parts. The control part can be 1–3 symbols, and corresponded to the Physical Downlink Control Channel (PDCCH). With the smallest bandwidth of 1.4 MHz the allocation range is from 2–4 symbols to ensure enough transmission capability for the PDCCH link adaptation in case cell edge users. The rest of the sub-frame is filled with data, which corresponds to the Physical Downlink Shared channel (PDSCH). The

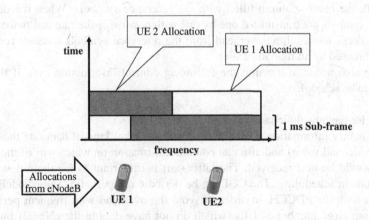

Figure 3.4 Uplink multiple access principle.

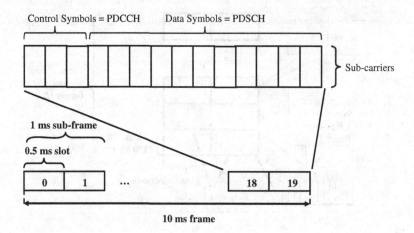

Figure 3.5   LTE FDD downlink frame structure.

allocation space for the PDCCH is dynamically signalled every sub-frame on the Physical Control Format Indicator Channel (PCFICH) which informs whether a single OFDMA symbol is needed for the PDCCH capacity needs or if two or three symbols are used.

Besides the earlier mentioned downlink physical channels, there is the physical HARQ Indicator Channel (PHICH) which informs the UE whether a packet in the uplink has been correctly received or not. Further there is the Physical Broadcast Channel (PBCH) carrying system information, or rather the Master Information Block (MIB) to indicate when the actual System Information Blocks (SIBs) are transmitted on PDSCH.

In the uplink direction the data is carried on the Physical Uplink Shared Channel (PUSCH). The PUSCH is constructed as shown in Figure 3.6. From the allocated uplink resource, two tables are filled with the data as shown in Figure 3.6 with the centre column filled with reference symbols while data and other control information filling the rest of the table. Then the columns are transmitted one by one with the cyclic prefix then being added after each column. When the data rate is doubled, the table per each slot has double amount of rows with the centre column filled with the reference symbols. When the data rate is doubled the symbols are transmitted one by one with now twice the rate and twice the bandwidth. This keeps the resulting overhead from the reference symbols constant regardless if the bandwidth used by an individual user.

There are also needs to transmit the following control information even if there is no uplink data to be scheduled:

- Feedback for the downlink packets received (ACK/NACK).
- Channel Quality Information (CQI), which indicates what kind of data rate the UE could receive (wideband value) and also can contain information on which part of the spectrum the data would be best received. The latter part is important to facilitate downlink frequency domain scheduling. The CQI can be periodic or aperiodic. The eNodeB can sent CQI request on the PDCCH, in order to avoid the overhead with frequent periodic CQI reports from large number of UEs (which do not have data in the eNodeB buffer to be transmitted).

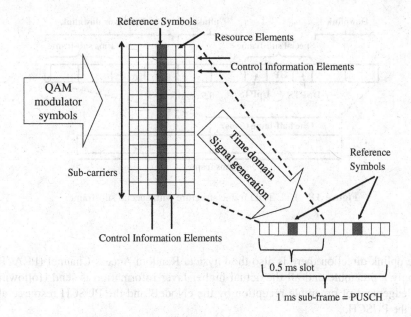

Figure 3.6   Uplink data transmission on PUSCH.

- Scheduling request, to give indication to the eNodeB that UE has uplink transmission needs. Further on the MAC layer there are more detailed reporting possibilities on the buffer status as well as on the uplink power resource availability.

When there is no PUSCH allocation, the signalling needs can be handled with the separate Physical Uplink Control Channel (PUCCH). For the use of PUCCH there is bandwidth reserved at the band edges since the standalone PUCCH is transmitted with the two 0.5 ms at the opposite end of the spectrum. To avoid unnecessary resource booking there is also CDM component to isolate different users using the same resource space (time and frequency) of PUCCH. An example of PUCCH allocation is shown in Figure 3.7.

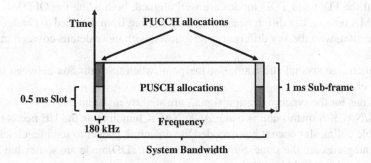

Figure 3.7   PUCCH resource allocation.

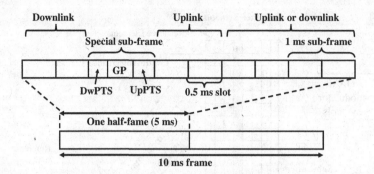

Figure 3.8   LTE TDD frame structure with special sub-frame.

In the uplink direction there is also the Physical Random Access Channel (PRACH) but this is only a preamble and all the actual higher layer information is send (following the acknowledgement of preamble reception by the eNodeB and the PUSCH resource allocation) on the PUSCH.

With the LTE TDD mode of operation (often referred to also as TD-LTE), the frame structure is otherwise the same but there is a special sub-frame when the transmission direction between downlink and uplink is changed, as shown in Figure 3.8. The physical layer parameters, including the bandwidths, are the same between TDD and FDD, with the intention in 3GPP to minimize the differences and use only specific solution for FDD and TDD when common approach could not be used (or would drop the performance significantly). Depending on the parameterization the special sub-frame occurs every 5 ms or then once every 10 ms frame if the transmission direction from downlink to uplink is only adjusted once per 10 ms. The selected uplink/downlink configuration impacts then the resulting maximum uplink and downlink data rates, with more downlink allocation the downlink peak data rate is higher but the uplink peak rate and capacity are reduced respectively. The sub-frames following the special sub-frame can be configured to be uplink or downlink but the first sub-frame after the special sub-frame is always allocated to uplink direction. This is needed to ensure minimum uplink capacity as well as Random Access Channel (RACH) transmission instants.

In general the FDD and TDD modes are well aligned, both using the OFDMA downlink and SC-FDMA uplink. The differences are mainly coming from the need to change the transmission direction, with the key differences as follows, with more details covered in [1]:

- TDD contains the special sub-frame for the point when transmission between uplink and downlink changes.
- The locations for the synchronization signals are slightly different.
- The ACK/NACK transmission is using ACK/NACK bundling as the UE needs to wait for an available uplink slot once it has decoded the downlink packet to send feedback.
- The UE categories are the same, but the peak rates in TDD mode are somewhat lower and depending on the uplink/downlink configuration as part of the sub-frames are needed for the other transmission direction.

Table 3.2   Downlink transmit modes for single antenna and multi-antenna operation

Transmission modes in LTE Release 9

Transmission Modes	Designation	Comment
TM 1	Single transmit antenna	Single antenna port: port 0
TM 2	Transmit diversity	2 or 4 antennas
TM 3	Open-loop spatial multiplexing with cyclic delay diversity (CDD)	2 or 4 antennas
TM 4	Closed-loop spatial multiplexing	2 or 4 antennas
TM 5	Multi-user MIMO	2 or 4 antennas
TM 6	Closed-loop spatial multiplexing using a single transmission layer	1 layer (rank 1), 2 or 4 antennas
TM 7	Beamforming	Single antenna port, port 5 (virtual antenna port, actual antenna configuration depends on implementation)
TM 8	Dual-layer beamforming	Dual-layer transmission, antenna port 7 and/or 8

The use of Multiple Input Multiple Output (MIMO) operation is introduced in Release 8 already, with multiple possible Transmission Modes (TMs) in the downlink direction. An overview of the modes is given in Table 3.2, with the TM1 being the single antenna mode and TM2 the mode used with transmit diversity. The TM3, TM4 and TM6 represent the multi-stream transmission to a single user while the TM5 is used with Multi-User MIMO (MU-MIMO). With MU-MIMO each antenna is transmitting different data streams but they are intended for different UEs. TM7 supports the UE Specific Reference symbols (URS) for the beamforming while the Release 9 addition, TM8, combines the possibility to operate with URS but use also MIMO (multi-stream) transmission with beamforming antennas. At the moment the support of UE specific Reference Symbols with TM7 (or TM8) is only mandated for TDD mode of operation as the first phase FDD mode LTE networks use normal MIMO operation, such as TM3 for example. The MIMO operation is specified for two and four downlink transmit antennas but from the UE side the currently available number of antennas in the field is two, in-line with the relevant UE categories as discussed in Section 3.7.

In Release 10 LTE-Advanced added then TM9, as discussed in Chapter 6, with enables multi-stream beamforming with the new reference symbol structures in Release 10.

In the uplink direction only single antenna transmission is defined for the UE. The network has however possibility to configure two or more UEs to use the same uplink PRB allocation by given each UE a different reference signal sequence to be used. Whether another UE is transmitting or not simultaneously is not visible to the UE. This kind of MIMO operation does not increase the UE specific peak rate, only the total peak rate achievable in the cell. The challenge in this Virtual MIMO (or multi-user MIMO) is to find UEs that have both similar transmission needs and which from the frequency domain scheduling would fit to the same part of the spectrum while still offering sufficient separation between the channels the UEs are experiencing.

3.3   Architecture

The LTE overall architecture has adopted the principle of flat radio architecture, as compared
to the Release 6 HSPA architecture both radio and core side user plane handling can be
handled in a single element, with one box core and one box radio, as shown in Figure 3.9.
The key motivation is to ensure easy scalability to avoid having to make capacity upgrades to
multiple levels like with the Release 6 case when the traffic is increasing. There have been
also further developments on HSPA side in Release 7 to add another architecture option, as
shown in Chapter 14, but in LTE there is only the option 'all radio in the eNodeB' being
included in the specifications.

Also the control plane design in the LTE radio is based on the single element radio access
network principle, thus all the radio protocols terminate in eNodeB as shown in Figure 3.10.
The connection to the core network is handled by S1 interface, which is divided in two to
allow maximum scalability between the user plane and control plane processing needs. User
plane uses S1-U interface while the control information between eNodeB and MME is
carried on the S1-MME interface. For the support of Radio Resource Management (RRM)
there is the X2 interface between the LTE eNodeBs. The X2 is mainly for the control plane
purposes but also in connection with the mobility events it is used for temporary user data
forwarding as discussed in Section 3.4. From the network planning point of view the X2
interface is a logical interface instead of any direct physical connection between the
eNodeBs.

The overall architecture showing both LTE Radio Access Network (LTE RAN) and
Evolved Packet Core (EPC) is shown in Figure 3.11. The S1_U interface is providing the
user data between the eNodeB and core network gateways. The main functionalities of the
EPC elements are further addressed in Section 3.5.

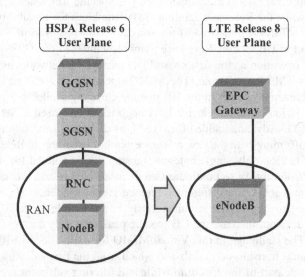

Figure 3.9   Evolution to the flat radio architecture with LTE.

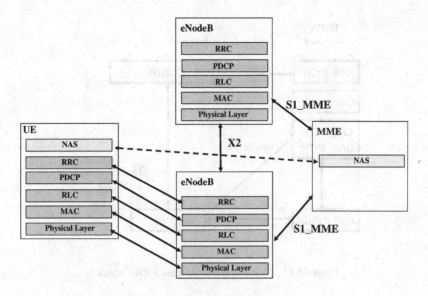

Figure 3.10   LTE radio protocol architecture.

3.4   Protocols

As visible in Figure 3.11, all the LTE user plane and control plane radio protocols terminate
in the eNodeB. On the control plane side the Radio Resource Control (RRC) protocol
provides the UE the RRC control messages which included configuring the connection
parameters, control of the mobility measurement and procedures, measurement reports from
the UE to the eNodeB, handover commands from the network to the UE and so on. Similar to

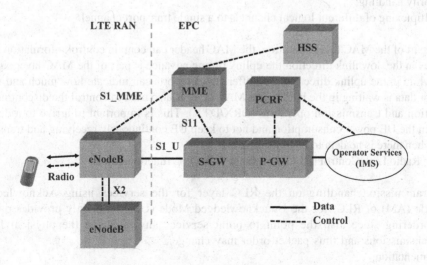

Figure 3.11   Overall architecture with LTE RAN and EPC.

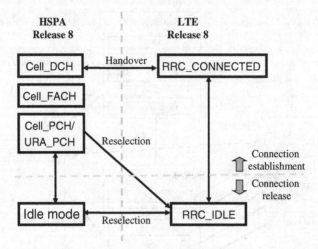

Figure 3.12   LTE and HSPA Release 8 RRC states.

the HSPA the same ASN.1 encoding is used with LTE, allowing extensions with newer 3GPP Releases to be implemented in a backwards compatible way. The key difference to HSPA is the definition of only two RRC states, as shown in Figure 3.12.

The Medium Access Control (MAC) is part of the layer 2 protocols with the key functional split shown in Figure 3.13 showing the data flow radio bearers all the way to the physical layer. The MAC has the following key functionalities:

• Physical layer retransmission handling;
• Scheduling;
• Priority handling;
• Multiplexing of different logical channels to a single transport channel.

As part of the MAC layer operation, the MAC header can contain control information also, such as in the downlink direction the uplink timing advance is part of the MAC layer signalling while in the uplink direction the buffer status reports can indicate how much and what kind of data is waiting in the buffer. The MAC layer is also used to control the discontinuous reception and transmission operation (DRX/DTX). This is important to arrive to a decent level in the UE power consumption and not to keep UE continuously receiving and transmitting when there is no data to be transmitted.

The Radio Link Control (RLC) covers the following functionalities:

• Retransmission handling on the RLC layer for the services using Acknowledged Mode (AM) of RLC, for the Unacknowledged Mode (UM) RLC only provides packet re-ordering since also the point to point services anyway have the physical layer retransmissions and thus packet order may change.
• Segmentation.
• Provisions of the logical channel to higher layers.

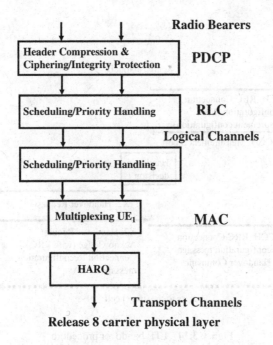

Figure 3.13   LTE Release 8 and 9 Layer 2 structure.

The Packet Data Convergence Protocol (PDCP) covers the following:

- Ciphering,
- Header compression,
- Integrity protection.

Normally all the data to be transmitted goes via the PDCP layer, the exception is the RRC signalling which is broadcasted to all UEs as there is no need for ciphering. The header compression is especially important for the operation of voice since the actual packet size is small and thus lot of capacity would be wasted without the header compression. The voice capabilities are fully completed in Release 9, which introduction of the solutions important such as support for the prioritized emergency calls as well as the necessary position location required in some of the markets, such as US. Further information on the LTE voice solutions can be found from [2].

The handling of UE mobility is one of the key procedures from the radio protocol point of view. The LTE UE is connected to one cell at the time only, and thus does not support macro-diversity as with WCDMA. The UE will search for the cells (for intra-frequency cells no neighbour list is needed) and after reporting on the Dedicated Control Channel (DCCH) which is sent in the uplink PUSCH, the eNodeB can make the handover decision as shown in Figure 3.14 to handover the UE to another cell. In this case, before the core network has rerouted the data path, the source eNodeB will use data forwarding to provide the data over the X2 interface to the target cell. This allows keeping the break in the data flow quite short

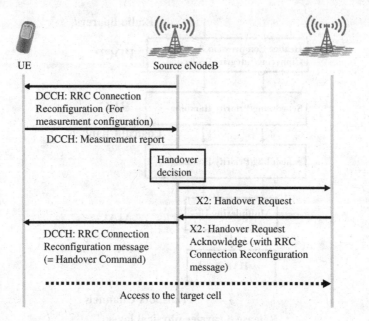

Figure 3.14   LTE handover procedure.

and also the overall handover procedure is very fast as core network can reach only once the handover is done.

3.5   EPC and IMS

In the core network side two key entities are separated. First of all there is the Evolved Packet Core (EPC) and then there is the IP Multimedia Sub-system (IMS), as shown in Figure 3.15. All user data goes through EPC and the data related to the operator IMS services (such Voice over LTE) goes also via IMS while traffic to the normal internet does not. The key EPC elements and functionalities are as follows:

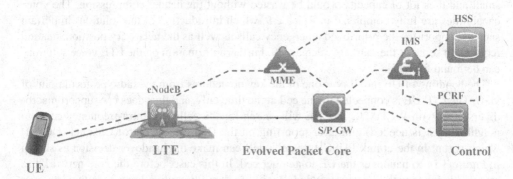

Figure 3.15   EPC and IMS.

- Mobility Management Entity (MME) is the main control element in EPC. It is contacted by the eNodeB for the UE authentication and MME obtains the necessary information from the registers about the subscription. The MME gets the subscription profile from the home network and based on that determines what kind of connections can be allowed for the UE, including parameters such as the maximum data rate.
- The MME is involved in the mobility managements as well. The MME will track where the UE is in the network and instructs the gateways where the data needs to be provided for the UE, updating this information in connection with the handover operation. In RRC_IDLE the MME follows in which tracking area the UE is in so that the network knows from which cell to page the UE in case incoming call.
- Serving Gateway (S-GW) key functionality is to handle the user data tunnelling based on the instructions received from the MME. When the UE moved in the network with active connection, the S-GW stays the same, like anchor point, and just routed the data to the correct eNodeB. It is possible however to have the UE changed from one S-GW to another. The S-GW receives the packets from the P-GW.
- Packet Data Network Gateway (P-GW) is the edge router between the between the LTE/EPC and external networks. The traffic gating and filtering functions desired are performed in the P-GW.
- The Policy and Charging Resource Function (PCRF) handles the Policy and Charging Control (PCC).

The IP Multimedia Sub-System (IMS) is the service layer on top of the IP connectivity layer provided by LTE and EPC. The IMS has been designed to work seamlessly with the 3GPP defined access networks. For the IMS services there are own-registration and session management procedures, with the SIP protocol in use for setting up and controlling the sessions. The IMS architecture is covered in more detail in [1] and references therein and with respect use of IMS for delivering Voice over LTE (VoLTE) is covered in details in [2]. The database, Home Subscriber Server (HSS), is used by IMS the same way as EPC/LTE uses it for obtaining the subscriber data.

3.6   UE Capability and Differences in Release 8 and 9

In Release 8 and 9 there are 5 UE categories defined. The UE does not signal individual capabilities but is according to one of the Release 8 (or 9) 5 UE categories. The data rates start from 10 Mbps downlink and 5 Mbps uplink with category 1, but the first phase UEs in the market are according to category 3 with 100 Mbps downlink peak data rate (with FDD mode) and 50 Mbps uplink when network has 20 MHz carrier available. The downlink MIMO capability is part of the all LTE UEs, except the category 1 devices, as shown in Table 3.3. The same UE categories are valid in both Release 8 and 9. The Release 9 enhancements did not impact the peak data rate of the UEs, but were mostly focused on enhancements to enable voice. Besides the UE capabilities captured in [3], on the smallest feature level the Feature Group Indicators (FGI) in [4] also informs the network whether UE has been tested against the network with particular a feature, such as certain channel quality feedback modes and thus whether the related feature could be activated in the UE. This allows introduction of the UEs in the market reflecting the features actually supported in the networks and facilitates then later feature introduction in the networks. Example of such a feature is

Table 3.3 Release 8 and 9 UE categories

UE Category	Class 1	Class 2	Class 3	Class 4	Class 5
Peak rate DL/UL	10/5 Mbps	50/25 Mbps	100/50 Mbps	150/50 Mbps	300/75 Mbps
RF bandwidth	20 MHz	20 MHz	20 MHz	20 MHz	20 MHz
Modulation DL	64QAM	64QAM	64QAM	64QAM	64QAM
Modulation UL	16QAM	16QAM	16QAM	16QAM	64QAM
Rx diversity	Yes	Yes	Yes	Yes	Yes
BTS tx diversity	1–4 tx	1–4 tx	1–4 tx	1–4 tx	1–4 tx
MIMO DL	Optional	2 × 2	2 × 2	2 × 2	4 × 4

Semi-Persistent Scheduling (SPS) which is reserving resources for multiple transmission occasions (repetitive). Since the SPS is intended for reducing the signalling load when majority, if not all, UEs are using voice, it has not been implemented in the networks widely and thus the UE may indicate (when connecting to the network) that the SPS has not been tested thus the eNodeB knows not to active the SPS for the given UE. Once the testing capabilities are improved, then the specifications may be updated to mandate setting a particular FGI bit to true after certain version of the given Release, visible in the annex of [4]. For some of the features, the FGI bit indication may be different between FDD and TDD modes of operation, as visible from [4] from the March 2012 version onwards.

The key differences in Release 9 compared to the Release 8 are as follows:

- Introduction of the earlier mentioned TM8 to facilitate simultaneous use of multi-stream MIMO transmission together with beamforming antennas.
- Introduction of the protocol support for the prioritized IMS emergency call.
- Introduction of the UE based position location solution, based on the UE Observed Time Difference Of Arrival (OTDOA) to provide further method for the UE position location than just satellite based solutions.
- Introduction of additional temporary solutions for CS fallback based voice provisioning, which is the intermediate solution before faster solutions such as PS handover and eventually Voice over LTE (VoLTE) are implemented.
- Introduction of the Multimedia Broadcast Multicast System support in the LTE to allow the use of LTE for broadcast type of services in a Single Frequency Network (SFN) configuration.

Thus based on these differences the Release 9 performance is not much different from Release 8, especially if no beamforming antennas are used. However, the emergency call support, as well as the additional position location solution, are both important enablers for the LTE based voice solution, as addressed in more detail in [2].

3.7 Conclusions

In the previous sections, we have covered the LTE Release 8 and 9 key principles. The principles introduced are mostly valid with the LTE-Advanced, with the Release 10 enhancements on top of this to improve the performance and data rate capabilities. On the architecture side the only new element part of Release 10 is the relay, otherwise the architecture is unchanged, and on the protocol side the changes are in many cases limited to the introduction of the necessary parameterization of the new Release 10 features introduced. LTE Release 8 and 9 are already an excellent basis for the network deployment. The focus in Release 9 enhancements was especially in enabling LTE voice support with the Release 10 and beyond LTE-Advanced then showing the direction how to improve the system once the data transmission needs are further evolved as covered in the following chapters.

References

1. Holma, H. and Toskala, A. (2011) *LTE for UMTS*, 2nd edn, John Wiley & Sons, Ltd, Chichester.
2. Poikselkä, M. *et al.* (2012) *Voice Over LTE, VoLTE*, John Wiley & Sons, Ltd, Chichester.
3. 3GPP Technical Specification, TS 36.306 (December 2011), LTE UE capabilities, version 9.6.0.
4. 3GPP Technical Specification, TS 36.331 (March 2012), LTE Radio Resource Control, version 9.10.0.

4

Downlink Carrier Aggregation

Mieszko Chmiel and Antti Toskala

4.1 Introduction

This chapter presents the LTE downlink carrier aggregation, part of Release 10 LTE specifications. First, the principle of carrier aggregation is introduced with the focus on downlink carrier aggregation. Following the discussion on the carrier aggregation principles, the detailed impact to the protocols, procedures and physical layer is covered. This chapter then concludes with the downlink carrier aggregation performance investigation as well as with the overview of the downlink band combinations worked on in 3GPP radio standardization.

4.2 Carrier Aggregation Principle

The Release 8 LTE downlink carrier maximum bandwidth is 20 MHz. This is well-suited to the most of the frequency bands for LTE deployment as the continuous allocation per operator in a given frequency band rarely exceeds 20 MHz. When seeking to introduce the data rate with larger bandwidth, one could have considered larger bandwidths but a lack of continuous spectrum would have limited the usability to a few frequency bands only. The solution was to use carrier aggregation where multiple carriers of 20 MHz (or less) would be aggregated for the same UE. The UE would be receiving carriers at the same time using of multiple frequency bands simultaneously. This kind of carrier aggregation where the carriers are on different frequency bands is considered inter-band carrier aggregation. Inter-band carrier aggregation of two cells of 10 MHz each is also an attractive use case from the operator point of view. It allows the operator to achieve similar UE peak data rates as in case of a single 20 MHz carrier but with fragmented spectrum. If an operator has room for more than 20 MHz in the same frequency band, then intra-band carrier aggregation can be also used, as shown in Figure 4.1. Further motivation with the existing 20 MHz bandwidth was to maintain the backwards compatibility with the earlier LTE Releases. An existing Release 8 UE may access the network using a single carrier only, while a Release 10 carrier aggregation

LTE-Advanced: 3GPP Solution for IMT-Advanced, First Edition. Edited by Harri Holma and Antti Toskala.
© 2012 John Wiley & Sons, Ltd. Published 2012 by John Wiley & Sons, Ltd.

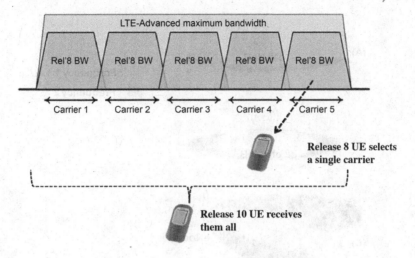

Figure 4.1 Basic carrier aggregation principle.

capable UE may then use more than one carrier without any impacts to the existing Release 8 or 9 terminals without carrier aggregation capability, as shown in Figure 4.2.

Each individual carrier in Release 10 handles all the Release 8 functions necessary for a Release 8 UE. This includes primary and secondary synchronization signals (PSS and SSS) as well as the Broadcast Channel and common reference symbols as covered in Chapter 3.

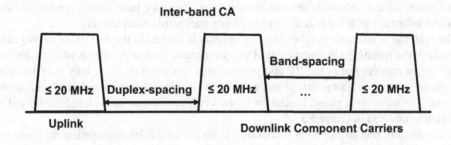

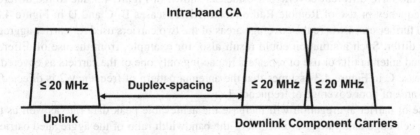

Figure 4.2 LTE downlink intra- and inter-band carrier aggregation principles.

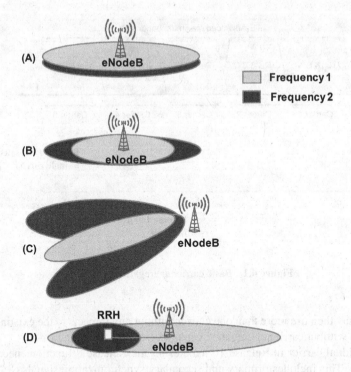

Figure 4.3   Different carrier aggregation deployment scenarios.

This creates some overhead when two carriers from the same base station are transmitting the same information, but was necessary to ensure backwards compatibility.

The both intra- and inter-band carrier aggregation is limited in the first phase to two carriers to limit the resulting UE complexity. The specification design principles would allow use of up to five carriers but in the RF and performance specification work only two downlink carriers are considered in the first phase, as covered later in connection with the band combinations. The same first phase limitation of two carriers is also considered in the uplink direction as covered in Chapter 5.

There are different deployment scenarios for the use of carrier aggregation, as illustrated in Figure 4.3. The aggregated carriers could be either with equal coverage or then the other carrier could have different coverage. Besides as shown in Figure 4.3 due to the different carrier frequency or use of Remote Radio Head (RRH) (cases B, C and D in Figure 4.3), there also further cases where the coverage areas of the two carriers used for carrier aggregation may differ. Such a situation could result also, for example, from the use of different antenna and antenna tilts or use of repeaters impacting only one of the carriers as covered in [1]. The case C in Figure 4.3 assumes that the antenna pattern of frequency 2 is directed to reach the area of poor coverage of frequency 1.

The use of carrier aggregation will increase the achievable peak data rate as well as the downlink cell edge data rate as a function of the bandwidth ratio of the aggregated carriers. Use of two 20 MHz carriers will prove double the peak data rate compared to the Release 8 case. If the extra carrier is smaller than the single carrier used, then the relative peak data increase is obviously less. For the cell edge performance the frequency band in use naturally will make a difference as aggregating a high (above 2 GHz) and low frequency band naturally contributes less to the improvement of the cell edge performance defined by the low band. However, the dynamic nature of the scheduler operation allows maximizing the use of the higher band. This leaves more of the lower frequency band capacity to those users that are not reachable with the higher frequency, as discussed more in connection with the carrier aggregation performance.

4.3 Protocol Impact from Carrier Aggregation

In this section, we discuss the impact on the LTE protocol structure and key procedures such as mobility from carrier aggregation.

The use of carrier aggregation is purely internal to the eNodeB. 3GPP specifications assume both carriers belong to the same eNodeB. In the user plane protocol design, the use of carrier aggregation is not visible above the Medium Access Control (MAC) layer. The Radio Link Control layer provides the logical channels as in Release 8, and if the carrier aggregation is enabled, the MAC layer functionality will then split the data on multiple downlink carriers, often referred as Downlink Component Carriers (DCC) also. The MAC layer will notify the RLC layer about transmission opportunities considering all carriers so that RLC PDUs can be formed accordingly, which is facilitated by the co-located RLC and MAC functionalities in eNodeB as shown in Chapter 3. Each of the carriers has their own physical layer retransmission handling as shown in Figure 4.4. There are separate per carrier HARQ entities, all HARQ entities of a UE use the same C-RNTI for scheduling because there is a single RRC connection per UE.

The scheduler functionality in the MAC layer will determine which of the component carriers should be used. This is done based on the carrier specific quality feedback (as discussed

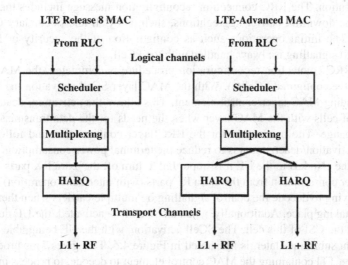

Figure 4.4 MAC structure with carrier aggregation.

later in connection with the physical layer impacts) as well as other available parameters, including the load on each of the carriers. Figure 4.2 shows one common scheduler for aggregated cells; however, it can be noted that the implementation with separate per carrier/cell schedulers which coordinate their scheduling decisions can achieve similar performance as the common scheduler. Separate and coordinated schedulers are more attractive from the complexity and scalability point of view compared to the common scheduler. Processing on the layers above in the Radio Link Control (RLC) layer or in the Packet Data Convergence Protocol (PDCP) layer does not differ between the Release 10 carrier aggregation and Release 8 single carrier operation. The PDCP functionality such as ciphering or header compression are handled as in Release 8 and then either acknowledged or unacknowledged mode RLC is used for the data being provided to the MAC layer. Respectively the operation on each carrier, including physical layer and HARQ operation is independent. A transport block failed on one of the carriers needs to be retransmitted with the HARQ retransmissions on the same carrier and on the same respective HARQ process, in-line with Release 8 HARQ principle. A single transport block is scheduled per component carrier in one Transmission Time Interval (TTI) to the UE, in case MIMO is in use in this carrier then two transport blocks can be scheduled per TTI on the carrier. Subsequent aggregated carriers increase the maximum number of the UEs transport blocks per TTI accordingly.

There is a single RRC connection only per UE regardless of the number of carriers being used. The Primary Cell (PCell) is always active and may be removed or changed only with handover. The determination of the error conditions, such as radio link failure are all based on PCell as in earlier LTE Releases.

The signalling flow in Figure 4.5 indicates the default bearer set-up (assuming the RRC connection set-up procedure, as discussed in Chapter 3, has been completed earlier) and attach procedure. The default bearer is set to enable data connection between the UE and rest of the network, after which the carrier aggregation is configured (if available both in the UE and eNodeB) with the RRC connection reconfiguration message. With the default bearer creation the core network controls the data rate (and other QoS parameters) to be used with carrier aggregation. The RRC connection reconfiguration message includes the parameters needed for the downlink carrier aggregations, such as PCell and Secondary Cell (SCell) information. The initial operations such as configuration of the security or Non-Access Stratum (NAS) signalling is always handled with the PCell.

Once the RRC connection reconfiguration procedure is completed, the MAC layer can then activate the configured SCell(s). With the MAC layer SCell activation the UE receives the corresponding MAC layer control element. This allows the activating, deactivating and reactivating of cells with the MAC layer when the needs for the data transmission or radio conditions change. The motivation for the RRC layer configuration and followed by the MAC layer activation/deactivation is to reduce the terminal power consumption. When there is no data in the eNodeB to the UE it is important to turn off the extra RX parts. This avoids both the power consumption from the extra RF parts (with inter-band operation) and also the UE avoids having to decode the control signalling on multiple carriers when there is no data transmission taking place. Additionally, when an SCell is deactivated, the UE does not measure or report the CSI of this cell. The SCell activation, with the UE being able to start data reception eight subframes later, is illustrated in Figure 4.6. This gives 7 ms processing time after end of the TTI containing the MAC control element to decode, to process inside the UE and to activate the receiver RF parts.

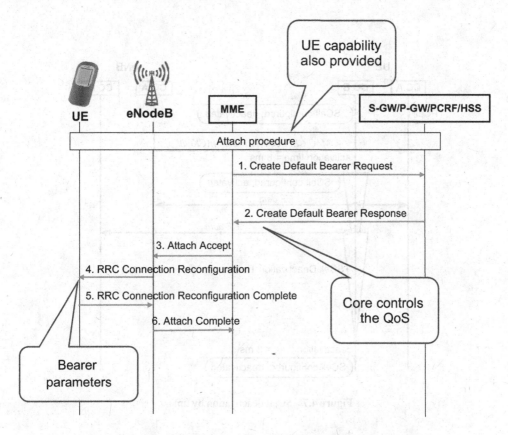

Figure 4.5   RRC connection set-up with carrier aggregation activation.

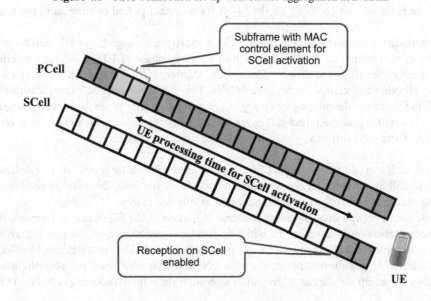

Figure 4.6   UE processing time for SCell activation.

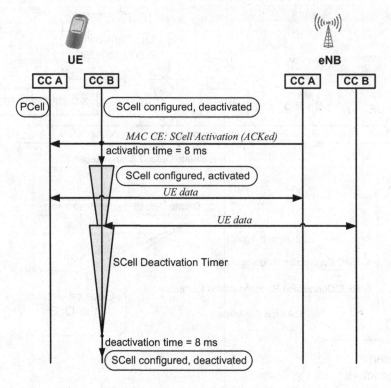

Figure 4.7 SCell deactivation by timer.

The SCell can also be deactivated by the configurable SCell deactivation timer in case there is no UE data activity on the SCell for a certain period of time as illustrated in Figure 4.7.

The mobility with carrier aggregation is kept mostly unchanged; the UE will base the mobility measurements on the PCell, which in the early phase of LTE-Advanced specification was also referred as the Primary Component Carrier, PCC. There is only one PCell for the UE; all other aggregated carriers are SCells. This limits the needed extra measurement capabilities from the downlink carrier aggregation capable UE. When handover is needed, the eNodeB will signal the target cell (over the X2 or via the core network) of the need for handover. There exist different possibilities for the handover handling:

- The eNodeB can add or reconfigure SCells to be used with the target PCell. The dedicated RRC signalling provides all necessary parameters in such a case, there isn't a need for the UE to decode the SIBs/System Information of SCells due to this.
- In case the target cell does not support carrier aggregation, the SCell can be removed with the reconfiguration in connection with the handover message as shown in Figure 4.8. Respectively when a UE is moved from a pre-Release 10 eNodeB to a Release 10 eNodeB with carrier aggregation support, the new eNodeB can then send the reconfiguration message to set up the carrier aggregation operation once the handover procedure is first completed.

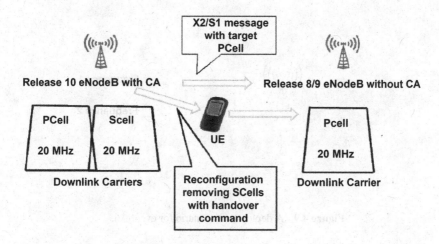

Figure 4.8 Mobility between Release 10 and 8 cells.

The Release 8 mobility events covered for the intra-LTE case in Chapter 3 are basically valid with the carrier aggregation as well. The eNodeB has to set a measurement object for the frequency of each component carrier in order to obtain measurement reports (once triggering conditions are fulfilled). The UE is also able to make measurements on the SCell frequency without measurement gaps, and these measurements can be used to determine if SCell is useful or whether handover should take place for another eNodeB. UE RRM measurements on the frequency of the configured, but not active, SCell are performed according to a configurable measurement cycle in order to reduce the UE measurement activity. The existing measurement cases can be used as follows, including one new one:

- Events A1 and A2, that is, a serving cell becoming better or worse than a threshold can be configured on all serving cells
- Events A3 and/or A5, that is, neighbouring cells becoming better (+ offset) than PCell or like with A5 the PCell below an absolute value and neighbouring cell better than an absolute value, can be configured on the frequency of PCell and SCell(s) (as shown in Figure 4.10 later).
- There is a new measurement event A6 introduced to the carrier aggregation needs. The new event A6 informs if an intra-frequency neighbouring cell becomes better (+ offset) than the current SCell. Setting a relative high offset requirement can be used here to avoid unnecessary measurement reports but again proper parameterization is also impacted by the PCell situation as well (such as expected relative coverage etc.). Event A6 can be particularly useful in the deployment scenario where the antennas of different aggregated carriers are pointed to different geographical areas as shown in Figure 4.9.

Figure 4.10 illustrates the measurement events in case of carrier aggregation that are reported relative to the PCell. The existing Release 8 inter-system measurements are unchanged.

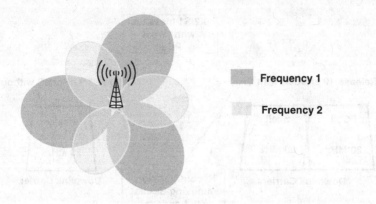

Figure 4.9   A deployment scenario for event A6.

In the idle mode the operation is unchanged to Release 8. A Release 10 UE does not behave any differently until it moves to the RRC-connected state and carrier aggregation is configured by the eNodeB (if supported by the eNodeB and if the UE capabilities indicate that the UE supports carrier aggregation in the given band combination).

4.4   Physical Layer Impact from Carrier Aggregation

In the physical layer the individual carrier is unchanged as far as a Release 8 UE is concerned. There are two modes which associate the Physical Downlink Control Channel (PDCCH) with the corresponding scheduled Physical Downlink Data Shared Channel (PDSCH) or the corresponding scheduled Physical Uplink Data Shared Channel (PUSCH):

1. The normal scheduling operation where the PDCCH and the corresponding data channel are transmitted on the same serving cell.
2. The cross carrier scheduling operation where the PDCCH can be transmitted on a different configurable serving cell compared to the corresponding data channel.

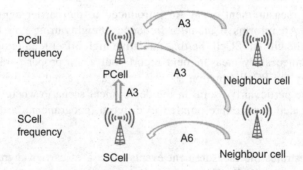

Figure 4.10   A new measurement event A6 with carrier aggregation.

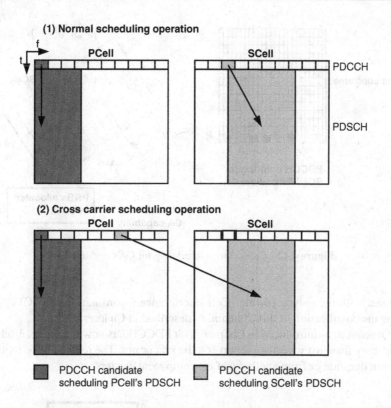

Figure 4.11 Normal and cross carrier scheduling operations.

These two modes are illustrated in Figure 4.11.

For the support of carrier aggregation with cross carrier scheduling, the Physical Downlink Control Channel (PDCCH) content is modified by adding to the PDCCH formats the Carrier Information Field (CIF) to allow indication for which of the downlink carriers the user resource allocation was intended. The general design principle is to enable up to five component carriers to be used but the real case for the first deployments (including performance requirements and band combinations) is two carriers. Use of cross carrier scheduling allows more dynamic use of the PDCCH resources as now also the PDCCH signalling capacity on both carriers is pooled, at least for the CA capable UEs as shown in Figure 4.12. One SCell is scheduled to the UE by PDCCH on one other serving cell only; however, PDCCH search spaces on this serving cell used for scheduling multiple cells to the UE can be shared.

The cross carrier scheduling is also beneficial with the operation with heterogeneous networks. The overlapping PDCCH regions may cause interference and thus by using the PDCCH on another frequency in the small cell, as shown in Figure 4.13, can be used to benefit fully from the use of the data part (where power level of the macro cell is smaller). The macro cell may avoid using the PDCCH on another frequency to ensure the micro/pico cell(s) can have interference free PDCCH on such a carrier. For the data part on PDSCH interference free position can be selected then with the normal frequency domain scheduling which does not address the PDCCH part being sent over the full carrier bandwidth. This can

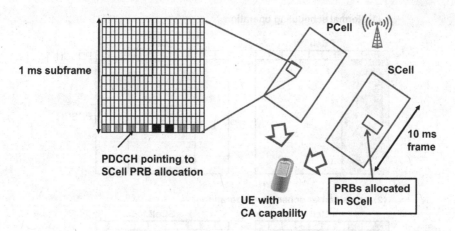

Figure 4.12 Cross carrier scheduling for CA capable UE.

be combined with the enhanced Inter-Cell Interference Coordination (eICIC) approaches addressing the coordination in time domain, as described in Chapter 8.

The CQI request, as introduced in Chapter 3, on PDCCH has now more than 1 bit to allow indicating more than just whether a report is to be sent or not. The report can be requested on the carrier in question or from another (set of) component carriers.

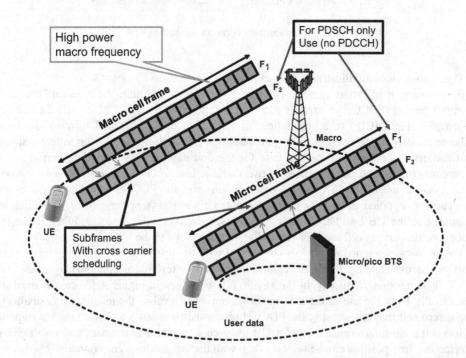

Figure 4.13 Use of cross carrier scheduling to facilitate heterogeneous network deployment.

The increase in the amount of the needed processing for the PDCCH blind decoding for a LTE capable UE has been limited by restricting the use of common search space on the PDCCH to the PCell only. This allows keeping the resulting processing increase on a reasonable level when adding more carriers to be received by the terminal.

In the uplink direction, also more signalling capacity is needed to cover the increased feedback needs due to the:

- Channel State Information (CSI) for more than one carrier, including Channel Quality information for each of the carriers as well as necessary MIMO related feedback. The CQI signalling for more than a single carrier is transmitted using PUSCH resource. If there are conflicts between the periodic CSI and aperiodic CSI, then the periodic one is dropped in the uplink direction. The CQI transmission following CQI request is taking place in connection with the uplink allocated resources on PUSCH. Moreover, periodic CSI reports of multiple cells are also supported on PUCCH by time multiplexing of the reports of different cells as configured by RRC.
- Physical layer retransmission related ACK/NACK feedback for the multiple downlink carriers to indicate whether a transport block was correctly received or not. In Release 8 ACK/NACK signalling was done for the transport blocks on a single carrier only.

The new PUCCH format 3 can carry 10 ACK/NACK bits (20 ACK/NACK bits for TDD), to enable sufficient ACK/NACK signalling capacity for five downlink component carriers. Furthermore, PUCCH format 1 b with channel selection is used to feedback up to 4 ACK/NACK bits which is sufficient to support aggregation of two cells. PUCCH is transmitted on the uplink of the PCell only. The PUCCH is also used for periodic CQI feedback, while the aperiodic feedback triggered with the CQI request was transmitted on PUSCH.

Specifically for TDD, there is a natural restriction that the uplink/downlink subframe split for all cells needs to be the same. Otherwise the UE transmitter in another carrier would block the UE receiver on another carrier. This is necessary especially when operating in the same frequency band, as shown in Figure 4.14. For the inter-band cases then more flexibility

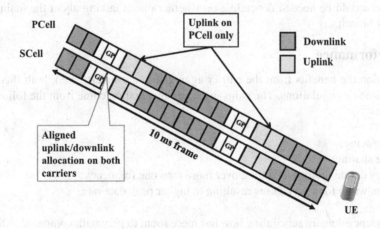

Figure 4.14 LTE TDD downlink carrier aggregation.

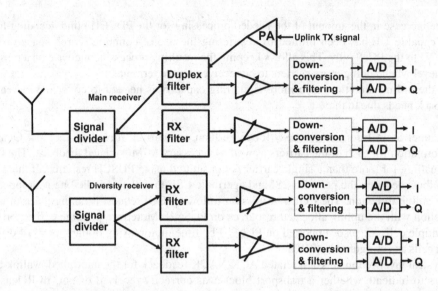

Figure 4.15   UE transceiver for inter-band downlink carrier aggregation case.

may be theoretically allowed, depending on the actual bands under consideration. In Release 10, however, the UE assumes all carriers have an identical uplink/downlink subframe split; the need for further flexibility will be addressed in Release 11 onwards. This restriction is not necessary a problem in reality as the first TDD carrier aggregation cases, as discussed in Section 4.6, are intra-band cases for the bands 38 or 41.

The implementation of carrier aggregation in most cases requires a separate receiver branch to be added, unless only intra-band adjacent carrier aggregation is considered which could be addressed with a single wideband RF front end. In Figure 4.15 the example receiver is shown with the band specific receiver parts behind the diversity antennas, assuming single uplink only thus duplex filter used in connection with one of the antennas only. In case of intra-band case (but non-adjacent carriers) only one band specific filter, either RX filter or duplex filter would be needed depending on whether one is talking about the main receiver or diversity branch.

4.5   Performance

In this section the benefits from the carrier aggregation are investigated, both theoretically and also based on simulations. The gains in carrier aggregation come from the following key implications:

- Load balancing;
- Resource sharing;
- Frequency domain joint scheduling over more than one frequency;
- More bandwidth for a given uses resulting to higher peak data rate.

The frequency domain scheduling now has more room to play with compared to the single carrier operation, as it can be based on the CSI/CQI selecting the resource blocks to be used

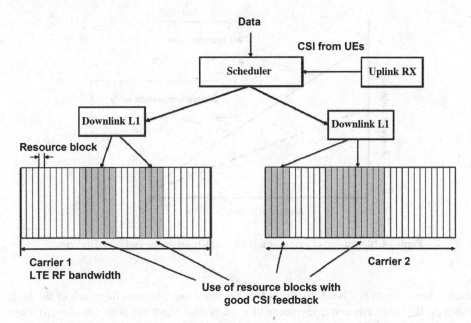

Figure 4.16 Frequency domain scheduling over two downlink carriers.

from the two different frequencies. This makes it more likely to find high SNR resource blocks with such a fading, and interference situations that allow use of higher order modulation, dual stream transmission and low channel coding overhead, thus maximizing the resulting system capacity. As shown in Figure 4.16, the eNodeB scheduler will select the resource blocks from two downlink carriers based on the UE feedback in the uplink direction.

The use of downlink carrier aggregation with high and low bands enables better use of the resources of both bands. With the higher frequency band there are users that with a single carrier operation should be operated only using lower frequency band to secure the connection maintenance. Now, when the Primary Cell (PCell) is on the lower frequency, the connection maintenance is secured but the scheduler can dynamically push part of the traffic via the higher frequency when indicated to be available by the CQI feedback. Thus the usable area for the higher frequency band is increased, as shown in Figure 4.17, leaving more of the

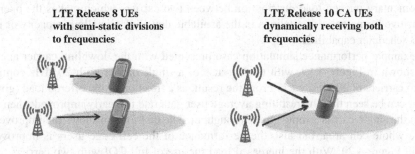

Figure 4.17 Dynamic frequency selection with downlink carrier aggregation.

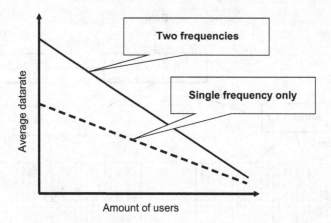

Figure 4.18 Impact of aggregating two frequencies on the end user data rate.

valuable lower frequency band resources free for those users beyond the reach of the higher frequency; the hysteresis and delay needed for the normal handover does not allow to it react to the instant changes in the fading, carrier load or interference. In general, also the amount of spectrum with lower frequency bands is less than with a higher frequency band, thus the use of carrier aggregation helps to make the LTE more attractive when not being limited only to the data rates offered by the 10 MHz LTE deployment in the lower frequency bands.

With the use of two high (or low) frequency bands, or in case of intra-band carrier aggregation, the two carriers are more balanced and will permit most of the users to use both of the carriers at all times (with intra-band carrier aggregation obviously there is no difference on the average path loss between the carriers). Two lots of available bandwidth for a given user directly give an increase in the average user data rate. The increase is better the less traffic and individual users there are on the available carriers. The availability of a very large number of users tends to provide a good scheduling possibility on any of the carriers but there is still a performance benefit for a carrier aggregation enabling device. In a less loaded network the average end user data rate roughly doubles with the use of two (equal bandwidth) carriers instead of only a single one, as shown in Figure 4.18.

The overall resource sharing over the two frequencies (or potentially more in the future) helps to maximize the achievable capacity. Using pooled resources avoids restrictions caused by a momentary uneven load distribution between two carriers which makes the operation less sensitive for potential limitations in the available downlink signalling capacity or in the eNodeB scheduler capabilities.

The example performance simulation case presented with the downlink carrier aggregation is shown in Figure 4.19, with the two use of a single carrier with 20 MHz compared with two carriers of 20 MHz each. From the result, as a function of the offered load (given in Mbps) it can be seen that the resulting average user data rate is clearly improved when compared to the single carrier average user throughput. The distribution shows the improvement over the whole cell area, but also the performance of the cell edge users is improved as shown in Figure 4.20. With the increased load the use of full CQI with two carriers comes close to the Release 8 CQI performance.

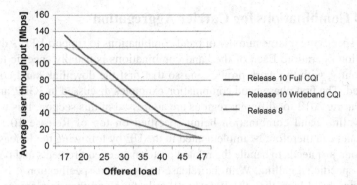

Figure 4.19 Downlink carrier aggregation average cell performance.

For the cell edge users better CQI is even more important as that allows the selection of the better frequency from the inter-cell interference point of view and within that, carries the least interfered-with resource blocks. With a very large number of users, eventually there is not that much benefit and the users at the cell edge may be configured back to a single carrier operation to minimize the amount of signalling needed.

The peak data rate impact can be calculated by simply adding the peak data rate on each of the carriers aggregated together. An LTE UE with two receiving antennas in Release 8 can support 150 Mbps on a single carrier, thus a Release 10 UE with the capability to aggregate two downlink carriers can then support a 300 Mbps peak data rate. With four-stream MIMO the single carrier data rate in Release 8 was 300 Mbps and when increased to eight-stream MIMO, up to 600 Mbps per carrier. Thus with a maximum of five carriers aggregated, the peak data rate would become 3 Gbps. The UE categories are addressed in further detail in Chapter 12.

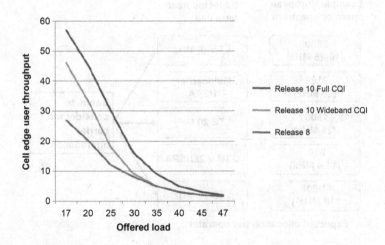

Figure 4.20 Downlink carrier aggregation cell edge performance.

4.6   Band Combinations for Carrier Aggregation

The 3GPP is specifying a large number of band combinations to support LTE-Advanced carrier aggregation operation. Each of the band combinations is usually specific to a specific region or country, with the work progressing so that first the downlink band combinations will be defined. The first phase band combination examples discussed in 3GPP, and targeting finalization during 2012 for different regions are addressed in this section. It is worth noting that the downlink band combination being specified on top of Release 10 are release-independent and can therefore be implemented in the UE by following the Release 10 specs: additionally one just needs to handle the related performance requirements and possible band combination specific signalling. With the release-independent specification it is possible to add a frequency band later than the Release 10 milestone and then implement Release 10 compatible carrier aggregation band combinations. The implementation is then done on top of Release 10 specifications, taking only the band specific signalling extensions and enhanced performance requirements into account in the implementation from the band-specific work done at a later stage.

In Europe, a typical operator has a spectrum available from different bands, including 800 MHz (the auctioning process is taking place in many countries soon, if not already), 900 MHz (from original GSM allocation, now partly for UMTS as well), 1800 MHz (GSM originally), 2100 MHz (original UMTS band) and 2600 MHz (intended for LTE) following the recent auctions. As shown in Figure 4.21 this suggests there would be 3 bands in the first phase to consider for the carrier aggregation use. There are other bands in Europe as well, such as the 3.5 GHz which is typically not necessarily owned by the same operator as the one having the bands in Figure 4.21.

For the European allocation the work is now focusing on the following band combinations:

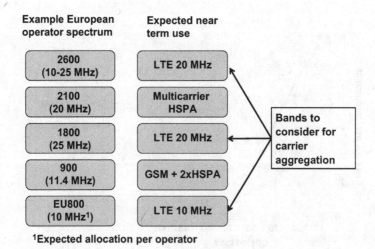

¹Expected allocation per operator

Figure 4.21   Example European operator band allocation.

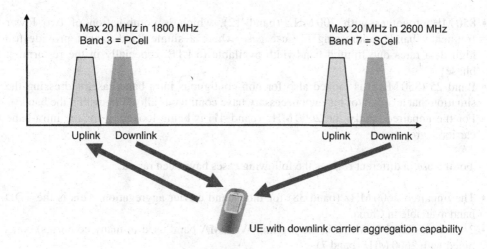

Figure 4.22 Downlink carrier aggregation with band 3 and band 7.

- 1800 MHz (band 3) combined with 2600 MHz (band 7). With the 2600 MHz there are typically 10–20 MHz or even a bit more available per operator. The 1800 MHz band (currently used for GSM mainly) is being considered as a suitable candidate band for re-farming to LTE and would form such a combination with 2600 MHz that would be widely usable across Europe. In the example shown in Figure 4.22, the PCell is on band 3 while only the band 7 downlink is received with the SCell. The setting could be other way around as well, with the PCell (with both uplink and downlink) on band 7. This band combination was introduced first as band 7 is already licensed in larger number of European countries and on that band there are several LTE networks already operational.
- 800 MHz (band 20) combined with 2600 MHz (band 7), addresses the restriction from the data rates with the expected 10 MHz available typically per operator in the most European markets where 800 MHz licensing has taken place (in some European countries also LTE networks are already running on band 20: for example, in Germany the auction was run in 2010 while in some other countries the auction is still to take place in 2012 and 2013 time-frame). Another set of combinations being considered with the band 20 is combination with 900 MHz, which is currently commonly used for GSM and 3G or with 1800 MHz currently used with GSM.

For the US (and for the countries with similar allocation) the first band combinations include:

- 1700/2100 (band 4) and 700 MHz (band 17), combining the coverage band at 700 MHz with a higher band frequency. The same 700 MHz is also considered with 1900 MHz (band 2) and with 850 MHz (band 5).
- The 1700/2100 (band 4) is also being looked with other 700 MHz bands (bands 13 and 12) as well as 850 MHz (band 5) and 2600 MHz (band 7).

- 850 MHz together with 700 MHz (band 12), which is combination of two lower frequency bands, addressing the use case where a single band cannot provide too high data rates due limited bandwidth available to LTE (especially in the re-farming phase).
- Band 25 (850 MHz) is looked also for non-contiguous intra-band case, addressing the situation that an operator may not necessary have continuous allocation within the band.
- For the unpaired bands, the 2600 MHz (band 41) is being looked at for the intra-band carrier aggregation.

For the use in different regions the following cases have been raised:

- The unpaired 2600 MHz (band 38) for intra-band carrier aggregation. This is the TDD band available in China.
- 2100 MHz (band 1, which is the original WCDMA band used in many countries) combined with 2600 MHz (band 7).
- 1800 MHz (band) combined with 850 MHz, is available in Korea.
- 2600 MHz (band 7) for intra-band carrier aggregation which is available in China and in some European cases also with more than 20 MHz allocation for each operator.
- For Japan the band 1 (2100 MHz) combined with bands 18, 19 or 21 and also band 11 combined with band 18.

In the first phase, the work for the band combination covers the case with two downlink and single uplink case, and the uplink will be then addressed once the key issues; such as insertion loss due to supporting an extra band, reference sensitivity, transmit power reduction and other key parameters for the band combinations in downlink, are first resolved during 2012.

The cases listed previously cover the combinations included in the 3GPP work program by the end of 2011. During 2012, 3GPP is expected to introduce further band combinations as the market needs them and interest develops, while at the same time the performance requirements and test cases for the previously-listed cases are being finalized.

4.7 Conclusions

In the previous sections, we have covered the LTE downlink carrier aggregation principles as defined in 3GPP Release 10 specifications. The downlink carrier aggregation is an attractive approach to improve not only the available peak data rates, but also the user capacity, as well as in the average, but also at the cell edge, without adding more transmission or receive antennas compared to the Release 8 LTE implementation. The use of single uplink in combination with two (or more) downlink carriers allows having minimum impact to the uplink link budget with only an extra switch needed in the RF path. The first phase band combinations with actual market interest (following the few introductory cases to set up to prime the procedure to define the necessary requirements) to be completed during 2012 enable going to peak downlink data rates in the order of 300 Mbps with a two antenna MIMO operation. With more bands and more antennas the theoretical data rates range up to 3 Gbps. The downlink carrier aggregation can clearly be expected to be the most popular feature in the practical network deployments due to the major benefits achievable with reasonable complexity. This development is also visible in the major interest from the LTE network operators for the band combinations for different frequency bands.

Reference

1. 3GPP Technical Specification TS 36.300 (December 2011) Evolved Universal Terrestrial Radio Access (E-UTRA) and Evolved Universal Terrestrial Radio Access Network (E-UTRAN); Overall description; Stage 2, Version 11.0.0.

5

Uplink Carrier Aggregation

Jari Lindholm, Claudio Rosa, Hua Wang and Antti Toskala

5.1 Introduction

This chapter presents the LTE uplink carrier aggregation principles. The uplink carrier aggregation is also part of the Release 10 LTE specifications, similar to the downlink carrier aggregation. First, the key principles of uplink carrier aggregation are introduced. This chapter then proceeds to the detailed impact on the protocols, procedures and physical layer. This chapter also covers physical layer changes in the uplink in the resource usage within a single carrier in Release 10 with the introduction of uplink multi-cluster scheduling. This chapter finally concludes with the uplink carrier aggregation performance investigation and the expected implementation impacts and challenges with the different uplink band combinations.

5.2 Uplink Carrier Aggregation Principle

The Release 8 uplink maximum bandwidth is 20 MHz and then a UE is transmitting that as contiguous resource block allocation with n times 180 kHz as discussed in Chapter 3. In Release 10 the uplink carrier aggregation principle is to enable the UE to transmit on multiple uplink carriers simultaneously, as shown in the Figure 5.1 for the FDD two uplink carriers case with a single downlink carrier. Depending on the spectrum allocation the uplink carrier aggregation use can be divided to three categories as follows:

1. Transmitting two or more adjacent carriers. This is typically not possible with the paired band allocations but more as the property of the unpaired spectrum allocations (and then naturally operating in TDD mode). There are some frequency bands having more than 20 MHz per operator, especially in the 3.5 GHz range, though there it remains most likely that TDD will be also used.

LTE-Advanced: 3GPP Solution for IMT-Advanced, First Edition. Edited by Harri Holma and Antti Toskala.
© 2012 John Wiley & Sons, Ltd. Published 2012 by John Wiley & Sons, Ltd.

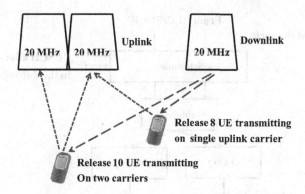

Figure 5.1   Uplink carrier aggregation principle.

2. Transmitting on two carriers in the same band but non-adjacent to each other. Similar to the downlink this is seen important for such a case when an operator holds non-contiguous spectrum allocation within the same band.
3. Transmitting two or more carriers with each of the carriers on different frequency band, with the inter-band uplink carrier aggregation. This is seen as the most typical uplink carrier aggregation use case on the basis of the requested band combinations in 3GPP.

Similar to the downlink carrier aggregation, a Release 8 UE can operate using any of the carriers (but only one of them at the time) whereas a Release 10 uplink carrier aggregation capable UE uses two uplink carriers with the uplink carrier aggregation.

On each individual uplink carrier a Release 10 carrier aggregation capable UE transmits in a similar way to the Release 8 carrier, but there are some differences in the resource usage in the uplink as now even within a single carrier case continuous spectrum allocation of the resources is not necessarily used (as discussed in Section 5.4 in more detail). Also, the uplink control channel (PUCCH) is only on the Primary Cell (PCell) frequency.

In the first phase the practical uplink carrier aggregation operation is limited to two uplink carriers, similar to the downlink carrier aggregation. The signalling in Release 10 specifications supports up to five aggregated uplink carriers but actual performance requirements work during 2012/2013 is only covering use of two uplink carriers. It remains to be seen whether the uplink inter-band carrier aggregation will be extended beyond two bands/carriers in the performance requirements in a later phase.

5.3   Protocol Impacts from Uplink Carrier Aggregation

The uplink carrier aggregation MAC layer operation is illustrated in Figure 5.2. The scheduling functionality is operating based on the PUSCH allocations received in the downlink direction. The eNodeB scheduler bases the uplink scheduling decisions as in Release 8 on the SRS sent in the uplink on each of the uplink carriers along with the MAC layer buffer status reports. Additionally the scheduler needs to consider available power resources in terms of whether the UE has enough transmission power to manage transmission on two carriers simultaneously or whether one should stay with a single carrier transmission only.

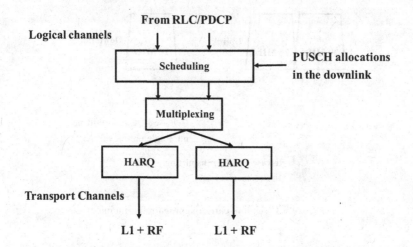

Figure 5.2   Uplink carrier aggregation protocol chain.

The UE MAC layer hides the carrier aggregation functionality from the RLC/PDCP layer and respectively in the uplink receiver side the eNodeB MAC layer combines the MAC PDUs to be a single stream for a given logical channel with reordering and RLC layer retransmissions as in Release 8, as described in Chapter 3.

The use of uplink carrier aggregation is not visible to the UE mobility measurements. The use of uplink carrier aggregation can continue in connection with the handover, if also the target cell supports it. Alternatively, it can be disabled with the RRC reconfiguration message in connection with the handover command.

During 3GPP Release 10 work a need was identified for multiple uplink timing advance values. These are needed if the network is using band-specific repeaters on one of the bands utilized for the uplink carrier aggregation with two (or more) uplinks in use. This was not included in Release 10 specifications but is being added to Release 11 specifications. With Release 11 specifications the eNodeB can give to the UE carrier (or rather frequency band) specific timing advance values to be used enabling the support of additional deployment scenarios such as the use of band specific repeaters as shown in Figure 5.3. In the environment in Figure 5.3 the upper band coverage is improved with a band specific repeater. Since a repeater has always some internal delay between reception and transmission of the amplified signal, different timing advance value is required compared to the signal which was propagated directly. There are potentially other use cases as well for multiple timing advance values if the different bands are served by different (and not co-located) antenna systems.

5.4   Physical Layer Impact from Uplink Carrier Aggregation

The PUCCH is always transmitted only in the PCell, regardless of whether there are more uplink component carriers configured for the UE. If there is uplink data scheduled in the PCell, then control is multiplexed on the PUSCH together with data in-line with the Release 8 uplink principles as was discussed in Chapter 3. The new element for the uplink in

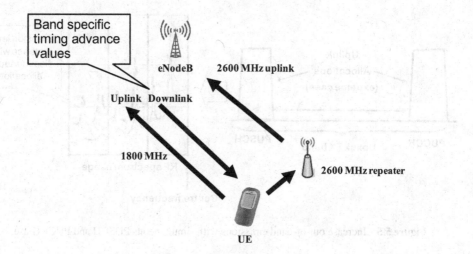

Figure 5.3 Use case for multiple uplink timing advance.

Release 10 (regardless whether uplink carrier aggregation is used or not) is the possibility to send simultaneous PUCCH and PUSCH, something which was not possible in Release 8 or 9. The principle of simultaneous PUCCH and PUSCH is shown in Figure 5.4.

The use of simultaneous PUSCH and PUCCH has the benefit of being able to set independently the power level between control and data since the data is protected by the physical layer retransmissions while the control signalling needs to go through in one go. With the simultaneous PUSCH and PUSCH transmission there are now more possibilities for the frequency specific scheduling in the uplink direction as previously only continuous allocation

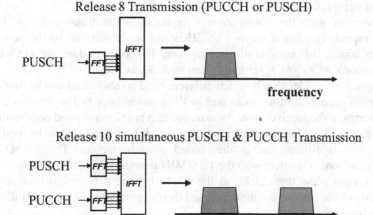

Figure 5.4 Simultaneous PUCCH and PUSCH.

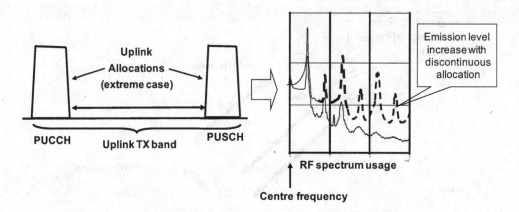

Figure 5.5 Increase out-of-band emissions with simultaneous PUSCH and PUCCH use.

was used. The use of simultaneous control and data (or two simultaneous data resource allocations) transmission, however, causes two parallel SC-FDMA transmissions to take place within the uplink allocations if a single uplink component carrier is used. This causes an increase in the emissions outside the carrier as shown in Figure 5.5.

The example shown in Figure 5.5 is only illustrative (and a rather extreme case with the use of both band edges) but studies have shown the need for Maximum Power Reduction (MPR). Thus use of simultaneous PUSCH and PUCCH (or PUSCH and PUSCH) should be avoided in a range critical situation at the cell edge due to reduced transmission power. The use of two parallel transmissions impacts spectrum emissions, spurious emissions (emissions outside own band) and, depending on the detailed case, there can be emissions due to the inter-modulation products of the two transmissions. In absolute terms the need for power reduction is in the order of 4–6 dB and in extreme cases even more. 3GPP is expected to determine the values to be used for each band or for each band combination in the case of uplink carrier aggregation.

When operating with the uplink carrier aggregation the transmission is otherwise unchanged, especially when a single PUSCH is in use. In Release 10 the new PUCCH format 3 was added. It is capable of carrying more bits as was addressed in Chapter 4 to cover the downlink ACK/NACK feedback from multiple carriers.

Another specific consideration is which different band combinations can be used. In some cases the uplink signal harmonics (second or third order) may fall in the downlink band causing problems in the receiver side. An example of a problematic band combination is the European 900 and 1800 MHz, as shown in Figure 5.6. If both bands would be used for LTE, then aggregating is difficult due to the second order harmonics. The 900 MHz uplink second order harmonics interfere with the 1800 MHz downlink reception. The case shown in Figure 5.6 may not occur that quickly as the 900 MHz band is being/has been refarmed to HSPA in many of the European countries (used then in parallel to GSM and HSPA) instead of refarming for LTE.

The actual transmitter implementation requires always two baseband chains. If we are only considering adjacent uplink carriers then a single wideband transmitter with a single power amplifier (PA) could be considered as shown in Figure 5.7.

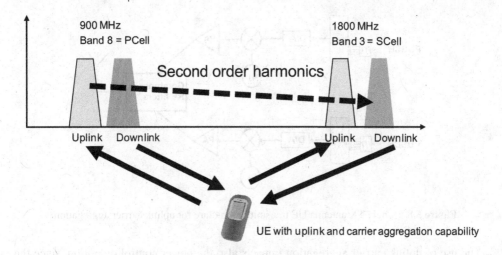

Figure 5.6   Example of uplink to downlink interference with carrier aggregation.

When using different bands for the different carriers then more components are needed. Figure 5.8 shows example of the UE with an added transmitter chain including two power amplifiers to enable transmission on two uplink carriers with non-adjacent carriers (either in the same or different bands). The total power shall not exceed the 23 dBm to stay within the UE power class. The Maximum Power Reductions (MPR) discussed earlier will cause the actual transmitted power to be smaller.

Combining the uplink inter-band carrier aggregation results to further complications if desired to be used together with the uplink MIMO, as discussed in Chapter 7, since two power amplifiers and related filters would then be needed per frequency with two antenna uplink MIMO operation and aggregating two uplink carriers.

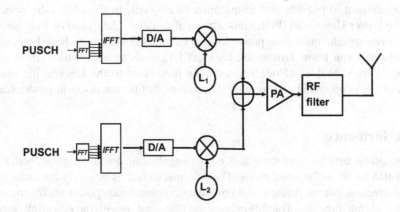

Figure 5.7   Uplink intra-band carrier aggregation with single PA.

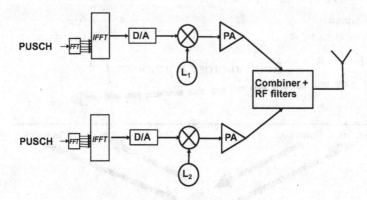

Figure 5.8   Single TX antenna UE transmitter structure for uplink carrier aggregation.

The use of uplink carrier aggregation impacts also the power control operation, since the UE power is now shared between more than one carrier. Transmission power on each carrier is controlled separately. Also as mentioned previously PUCCH power in the PCell is controlled independently from PUSCH in the same carrier. Independent power control of PUCCH and PUSCH in each carrier enables reuse of Release 8 power control formula also in case of carrier aggregation. However, because powers of PUCCH and PUSCH transmissions are independently controlled, the aggregated total output power of the UE could exceed the UE maximum power limit even if each individual PUCCH or PUSCH transmission is below maximum allowed transmit power. In this case powers of different uplink channels are scaled down so that PUCCH has the highest priority, PUSCH with Uplink Control Information (UCI) has the second highest priority and the remaining power is allocated to PUSCH without UCI.

As pointed out earlier, UE is allowed to reduce its maximum power limit if uplink allocation results in high spurious emissions or if the peak to average ratio of the transmitted signal is too high or if UE has some other radio system simultaneously transmitting so that total transmission power would exceed allowed limit. In order to schedule efficiently, eNodeB needs to know how close to the UE's maximum power limit it is operating. Power headroom reports are defined to provide that information to eNodeB. In Figure 5.9 the principle of Release 10 Power Headroom (PHR) reporting is illustrated. Compared to Release 8 PHR, also the carrier specific maximum power limit is included in the report. Based on that information, eNodeB can know how much back-off UE needs to have with current resource allocation. Also, it enables eNodeB to estimate how close to the UE specific maximum power limit UE is operating and to make allocations that do not result in power scaling in the UE.

5.5   Performance

In this section the benefits from the uplink carrier aggregation are investigated, both theoretically and also based on the simulations. The gain mechanism is basically the same as in the downlink direction, but the limited (and now shared) transmission power makes some differences in the uplink direction. The following areas (the same as with the downlink carrier) are improved with the uplink carrier aggregation, as also studied in [1]:

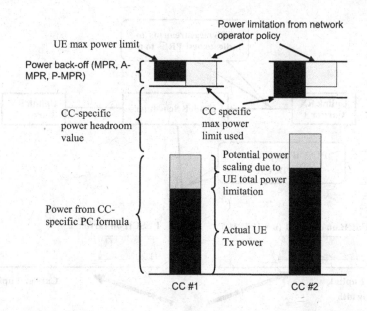

Figure 5.9 UE power setting and power headroom definition in case of carrier aggregation.

- Load balancing;
- Resource sharing;
- Frequency domain joint scheduling over more than one frequency;
- More bandwidth for a given user resulting in a higher peak data rate.

The load balancing and resource sharing is similar to the downlink. The eNodeB resources are now pooled on both carriers and load of the UEs with carrier aggregation can be shared on a 1 ms (TTI basis) on the two carriers. The use of Sounding Reference Signals (SRS) in the uplink now allows for uplink frequency domain scheduling, as shown in Figure 5.10. Alternatively the frequency domain scheduling can be done based on the interference measurements in the eNodeB. The eNodeB may configure the SRS patterns on each carrier as in earlier Releases as was discussed in Chapter 3, with the accuracy on each carrier depending then on the amount of SRS send on the uplink carriers in use.

The uplink peak rate is improved from the 50 Mbps (when using 16QAM) to 100 Mbps with two 20 MHz carriers, or from 75 Mbps to 150 Mbps if using 64QAM (but single antenna transmission). The use of 64QAM is not mandatory in the uplink direction in Release 10 either and can be seen from the UE categories addressed in Chapter 12. The uplink carrier aggregation can be also combined with uplink MIMO to reach even higher peak data rates, reaching to 300 Mbps with two 20 MHz carriers using 64QAM and uplink MIMO.

The use of frequency domain scheduling over two carriers should give (with low to medium load cases) similar performance gain as in the downlink direction; but as the total UE transmit power is limited, then the cell edge users do not normally get any gain by using carrier aggregation in uplink. Even if allocated on multiple component carriers, cell edge users do not have sufficient power to exploit the increased transmission bandwidth.

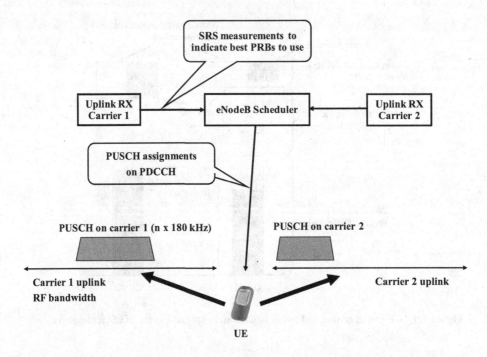

Figure 5.10   Uplink frequency domain scheduling based on SRS transmission.

With non-contiguous resource allocation in uplink the required power back-off in the UE power amplifier is increased, which consequently results in a reduction of the maximum transmission power. The exact power back-off depends on the specific uplink resource allocation and transmission mode and it is UE implementation specific. Therefore the Maximum Power Reduction (MPR) standardized in [2] and illustrated in Figure 5.11 is here used to model the required power back-off at the UE, though in practice the MPR only represents an upper bound for the power back-off. As can be observed, the MPR solely depends on the ratio between the allocated PRBs and the aggregated system transmission bandwidth.

In addition to carrier aggregation, multi-cluster transmission has also been introduced in Release 10 to further improve the spectral efficiency in uplink. With multi-cluster transmission, a UE can be allocated a maximum of two non-adjacent clusters within one carrier, so that higher scheduling flexibility can be achieved compared to SC-FDMA.

With multi-cluster scheduling there is also need for additional power back-off at the UE as in the case of carrier aggregation. Since the MPR standardized in [2] applies to multi-cluster transmissions independently whether clusters are within the same carrier or in different (contiguous) carriers, the power back-off model in Figure 5.11 can also be applied to the case with multi-cluster scheduling.

In Figure 5.12 the uplink multi-cluster performance is studied for the 16QAM operation with different number of UEs. With the 10 MHz case studied the gain from the use of two PUSCHs ranges between 10–15%. The gain will start to saturate as the number of UEs is increased since there are more UEs to choose from and the benefit from the multi-cluster operation is not that big anymore as more and more UEs are scheduled simultaneously. With

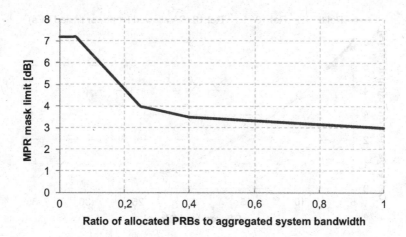

Figure 5.11 Standardized MPR used to model required power back-off in the UE with non-contiguous resource allocations in uplink [2].

a single UE case there is no benefit as the UE anyway can be scheduled over the whole bandwidth assuming there are enough power resources.

The results in Figure 5.12 look at the multi-cluster case in isolation from the other relevant feature such as Multi-User MIMO that is further addressed in connection with the multi-cluster allocation in Chapter 7.

Next, we look at the overall uplink performance of Release 10 (carrier aggregation and multi-cluster scheduling) compared to Release 8. The cell edge and average user throughput performance with two 20 MHz uplink carriers are presented in Figures 5.13 and 5.14, respectively, and assuming the power back-off model of Figure 5.11.

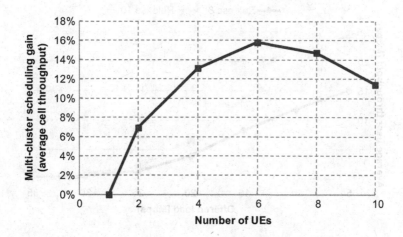

Figure 5.12   Multi-cluster scheduling gain.

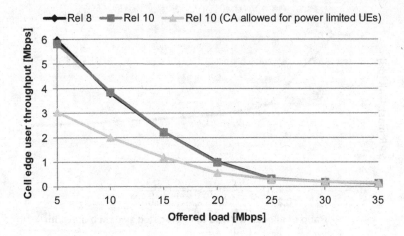

Figure 5.13 Cell edge user throughput with uplink carrier aggregation and dual-cluster scheduling compared to Release 8 performance.

For power-limited cell edge users transmitting at maximum transmission power the additional power back-off needed with carrier aggregation and multi-cluster scheduling actually results in a coverage loss compared to a Release 8 operation where the SC-FDMA properties of the transmitted signals are maintained. From the network point of view this clearly suggests avoiding using uplink carrier aggregation at the cell edge area, and to activate that only when a user is identified to be in a non-power limited case. As also studied in [3], distinguishing between power-limited and non-power-limited Release 10 terminals is the key to avoid throughput decrease at cell edge compared to Release 8 operation. Release 10 power-limited terminals are only assigned on one component carrier and scheduled with single-cluster transmission.

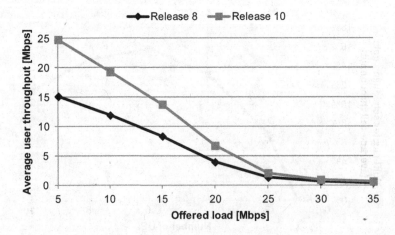

Figure 5.14 Average user throughput with uplink carrier aggregation and dual-cluster scheduling compared to Release 8 performance.

On the other hand, Release 10 terminals not operating close to their maximum transmission power can be assigned on both component carriers so that they can benefit from the advantages of carrier aggregation (and multi-cluster scheduling). With the high load situation the probability of a user getting scheduled on two uplink carriers is reduced, and so is the gain. Anyway, for lower offered loads and for a user close enough to the base station, the resulting uplink data rate can be twice as high compared to a single carrier data rate, though the observed gain at low load in practice lies in the range between 60–70% as visible in Figure 5.14.

5.6 Band Combinations for Carrier Aggregation

Basically the same band combinations as with the downlink carrier aggregation in Chapter 4 are considered for the uplink carrier aggregation. The use of two uplinks causes additional considerations for the interference and some band combinations may need to be used only with the downlink carrier aggregation. The 3GPP performance work will cover first the downlink band combinations with them being simpler than the uplink ones. The uplink will have some additional issues due to the following:

- It is not desirable to exceed the maximum transmission power thus the power on each of the uplinks on the average is reduced by 3 dB compared to the single carrier UE transmission power.
- The two transmitters may cause additional issues that impact out of band or spurious emissions and thus additional maximum power reduction is likely to be needed.
- In some cases the uplink transmission band may be too close to a downlink reception band. For example the use of the European 1800 MHz and 2100 MHz bands is problematic. The 2100 MHz band uplink transmission band edge at 1920 MHz is only 40 MHz apart from the downlink band edge at 1880 MHz for 1800 MHz band downlink and thus would cause TX noise to leak to the downlink reception band as shown in Figure 5.15. Such a band

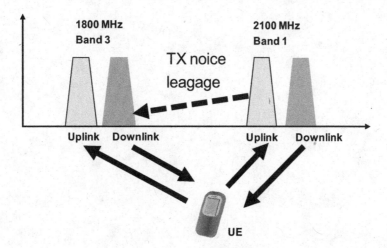

Figure 5.15 TX power leaking to the downlink receiver band.

combination has not been proposed so far for use in LTE carrier aggregation. Also the harmonics in the case of some bands may be a potential reason to limit the carrier aggregation to take place only in the downlink direction if the implementation would be too difficult. 3GPP has initiated work on some of the cases with known problems, with the work expected to be completed around mid-2013; for example, to consider 850 MHz together with 1800 MHz band [4].

5.7 Conclusions

In this chapter, we have covered the LTE uplink carrier aggregation principles as defined in 3GPP Release 10 specifications. The use of uplink carrier aggregation allows having higher uplink peak rates compared to Release 8 and reaching 100 Mbps in the uplink with single uplink transmit antenna and without having to use 64QAM modulation. Together with Release 10 uplink carrier aggregation there are extensions also for the single carrier uplink operation as the resource allocation in the uplink is now more flexible. The Release 10 specifications enable simultaneous use of two PUSCH allocations or alternative simultaneous PUSCH and PUCCH allocation with non-contiguous uplink resource allocation. The uplink carrier aggregation support is completed in Release 11 with the support of multiple uplink timing advance values to ensure operation with band specific repeater scenarios.

References

1. Wang, H., Rosa, C. and Pedersen, K. (2010) Performance of uplink carrier aggregation in LTE-advanced systems. Vehicular Technology Conference Fall (VTC 2010-Fall), in Proceedings, September 2010.
2. 3GPP TS 36.101 (December 2011) UE radio transmission and reception, version 10.5.0.
3. Wang., H., Rosa, C. and Pedersen, K. (2011) Uplink component carrier selection for LTE-advanced systems with carrier aggregation. International Conference on Communications (ICC 2011) Proceedings, May 2011.
4. 3GPP Tdoc RP-120364 (2012) Work Item Proposal: LTE Advanced Carrier Aggregation of Band 3 and Band 5 with 2UL, SK Telecom, 3GPP TSG RAN#55, February 28–March 2.

6

Downlink MIMO

Timo Lunttila, Peter Skov and Antti Toskala

6.1 Introduction

This chapter presents the LTE downlink MIMO enhancements included in Release 10 LTE specifications, as part of LTE-Advanced work to improve the performance and reach the LTE-Advanced requirements set in 3GPP, as well as the IMT-Advanced requirements by the ITU-R. First, an overview of the downlink MIMO enhancements over the Release 8 and 9 MIMO capabilities is presented. This chapter proceeds to cover the changes in the protocol side to enable the necessary signalling and then the physical layer impacts are presented. This chapter concludes with the downlink MIMO performance investigation and the expected implementation impacts.

6.2 Downlink MIMO Enhancements Overview

Downlink MIMO operation in LTE is available already in Release 8 LTE specifications and widely used in the first phase LTE networks rolled out based on the principles shown in Chapter 3. Release 10 provides several new enhancements which can both improve the performance and further boost the achievable peak data rates in the downlink direction. The following key new areas are introduced on top of the Release 8 and 9 MIMO capabilities as shown in Figure 6.1.

- Extension of the MIMO supports up to eight antennas with up to eight parallel data streams if eight transmitter and eight receiver antennas available (8×8 MIMO).
- Transmission with two, four or eight antennas and deriving the Precoding Matrix Indication (PMI)-related feedback based on the Channel State Information Reference Signals (CSI-RS)
- MIMO transmission with the use of UE-specific demodulation Reference Signal (URS) which are sent in the same physical resource blocks than the data for the given user.

LTE-Advanced: 3GPP Solution for IMT-Advanced, First Edition. Edited by Harri Holma and Antti Toskala.
© 2012 John Wiley & Sons, Ltd. Published 2012 by John Wiley & Sons, Ltd.

LTE Release 8	**LTE Release 9**	**LTE Release 10**
SU-MIMO 2x2, 4x2 and 4x4 based on Common Reference Signals (CRS). Single stream 8TX beamforming with user specific Demodulation Reference Signals (DM-RS) Limited MU-MIMO.	MIMO operation enabled together with beamforming Flexible MU-MIMO	Support for SU-MIMO with up to 8 layers (8x8). Precoding Matrix Indication (PMI) based on Channel Sate Information Reference Symbols (CSI-RS) up to 8 TX Single port CRS operation MU-MIMO enhancements.

Figure 6.1 Downlink MIMO development in different releases.

• Optimization of the MIMO feedback, including the feedback for both Multi-user MIMO (MU-MIMO) and single user MIMO (SU-MIMO).

The design criteria of the enhancements was, on the other hand, to ensure backwards compatibility with the Release 8 and 9 MIMO operation, while also reducing the impact to the system performance when supporting more than two antennas for MIMO transmission with the introduction of the solutions that minimize the extra reference signal overhead when having up to eight MIMO transmission streams in use. With the new structures, the overhead can be scaled depending on the amount of MIMO stream used to avoid capacity and performance degradation for the Release 8 and 9 terminals as well.

6.3   Protocol Impact from Downlink MIMO Enhancements

The downlink MIMO operation is visible mainly in the physical layer, however, the MAC layer scheduler is takes part in deciding the connection with the feedback received in the uplink direction, when a user should be scheduled, and what kind of rank (number of parallel MIMO streams) should be used, or whether the conditions would be proper for a Multiuser MIMO (MU-MIMO) operation. MAC layer operation in connection with downlink MIMO is illustrated in Figure 6.2.

The UE will derive the feedback from the CSI-RS, which is discussed more in Section 6.4. Then the eNodeB scheduler will determine which user to schedule based on the feedback received as well as the user priority, buffer status and other vendor-specific criteria. If multiple Release 10 features are available, such as carrier aggregation, the eNodeB scheduler needs to consider the situation across multiple carriers where a particular user is to be scheduled and with what kind of MIMO transmission is to be used.

There are several new parameters to configure in the RRC signalling for Release 10 downlink MIMO operation, including the CSI-RS configuration and what kind of feedback to provide. For the mobility measurements and actual procedures there are no changes due to the introduction of MIMO, as the mobility measurements are still assumed to be done based on the Release 8 CRS transmission.

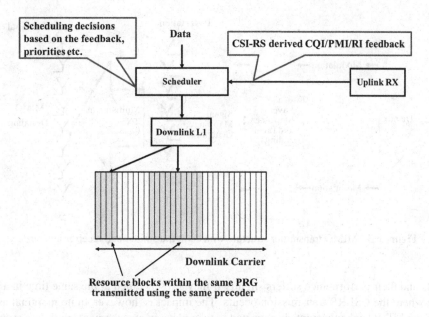

Figure 6.2 Downlink MIMO feedback for an eNodeB scheduler.

6.4 Physical Layer Impact from Downlink MIMO

The enhancing of the downlink multi-antenna capabilities mainly impacts the physical layer in the 3GPP specifications. The addition of the eight parallel streams increases the momentary peak data rate and requires these eight parallel data streams to be transmitted and received using eight antennas. Obviously, the eight stream transmission is not expected to be used in the very near future, simply due to the added complexity of supporting eight transmitter antennas in the eNodeB and eight receiver antennas in the UE side. The amount of actual code words to be transmitted has not been increased from two code words, even if up to eight layers are being used; this allows one to minimize the feedback needed. The possibility of going up to an eight -layer transmission is shown in Figure 6.3; new reference signal solutions and improved feedback form the new Transmission Mode 9 (TM9) on top of the eight modes defined in Release 8 and 9 as listed in Chapter 3.

The Channel State Information Reference Signals (CSI-RS) enable the use of terminal feedback to cover for the operation with up to eight antennas without having to send the Common Reference Signals (CRS) all the time for that purpose from four or eight antennas. The CSI-RS is intended for the feedback generation, thus it can be spaced more sparsely as reference symbols are needed for data detection. Thus, in a case a UE is in such a situation that MIMO operation with higher orders of MIMO (more than two streams) is not possible, only a very limited overhead is generated from the presence of the CSI-RS, compared to the situation that the full CRS would need to be sent from four or more antennas. The density of them varies, as in the time domain they can be parameterized to be send with a rate of 5–80 ms and they use only a single resource element per antenna port in each physical resource block. Release 8 or 9 terminals do not take advantage of the availability of these

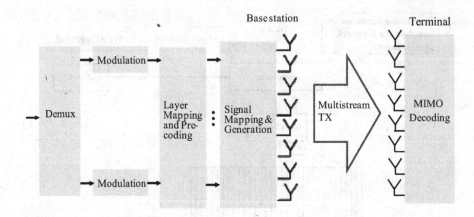

Figure 6.3   MIMO transmitter structure with eight transmitter and receiver antennas.

signals and their performance suffers slightly if they are scheduled at the same time in a sub-frame where the CSI-RS transmission occurs. The impact is, however, quite marginal as the use of the MIMO enhancements is expected to take place in environments with less mobility and thus, with the occurrence rate of 20 ms, only 5% of the sub-frames experience the slight degradation due to these extra symbols on top of the data.

An example of a CSI-RS configuration is shown in Figure 6.4 for the case of a legacy UE receiving the data from two antenna ports. With four or eight antennas in use then more resource elements are occupied for the CSI-RS during the sub-frame then being transmitted. The detailed parameterization may vary as there are different options available depending on whether one is using TDD or FDD or what type of cyclic prefix is being used.

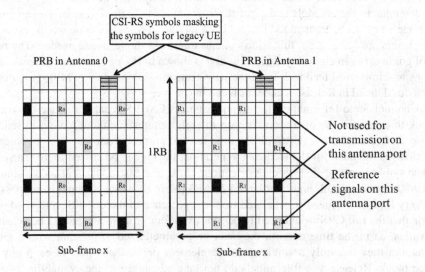

Figure 6.4   CSI-RS in case of two antenna ports.

As mentioned previously, there is effectively only one resource element allocated for one CSI-RS port in one sub-frame. To still maintain good quality for CSI-RS based measurements, support for higher reuse factors than 1 on the resource elements carrying CSI-RS are added. This means that the eNodeB can configure both a CSI-RS resource for measurement and also a set of resources where neither PDSCH nor CSI-RS is transmitted; the latter configuration is also referred to as the zero-power CSI-RS configuration. With this capability it is possible to achieve higher reuse factors on CSI-RS for example reuse three and thus improve the measurement quality. This could be relevant for situations where UE feedback modes are using sub-band based measurements where only REs from a few PRBs can be used. Forward compatibility to future releases where CSI measurements in neighbour cells might be supported is another reason for introducing the zero-power CSI-RS configuration.

When the UE is measuring CQI it is not only important to estimate the UE's own signal but also interference from other cells. With the introduction of CSI-RS there was some discussion on whether more detailed descriptions about how the UE measures the interference were needed. Assuming, for example, that eNodeB has configured CSI-RS with reuse 1, then interference measurements based on the CSI-RS would not be able to distinguish situations with high and low system load as the PDSCH would not interfere with CSI-RS. To avoid these issues it was decided to stick to the assumption of interference measurements based on Common Reference Signal (CRS) as used in earlier releases. Here, load awareness is always insured as the frequency shift of CRS positions is hardwired to the cell ID.

The UE specific Reference Signal (URS), (also known as Demodulation Reference Signals: DM-RS) are sent with the same precoding as the data itself, and in the same physical resource blocks as the data. For the resource blocks allocated to the Release 8 devices the URS is not used as those devices rely on the use of common reference signal and the precoding applied is signalled separately. An example sub-frame with URS is shown in Figure 6.5, assuming such a sub-frame where no CSI-RS transmission takes place, the URS is provided for a two-antenna case in the example with up to eight ports, double the amount (from three to eight layers) of resource elements is needed for the URS use. The use of URS causes extra overhead compared to the Release 8 as the CRS needs anyway to be retained for other use, including the mobility measurements by the terminals not having allocation, legacy or Release 10 UEs.

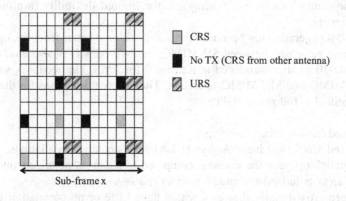

	CRS
	No TX (CRS from other antenna)
	URS

Sub-frame x

Figure 6.5   Release 10 URS with a two-antenna case for one antenna.

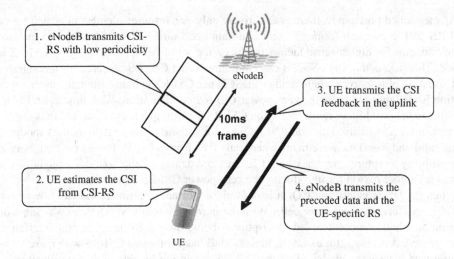

Figure 6.6 Downlink MIMO operation procedure.

The physical layer operation of the downlink MIMO is illustrated in Figure 6.6. The eNodeB transmits the CSI-RS with low periodicity (with the periodicity signalled to the UEs). The UE estimates from the CSI-RS received the CQI/PMI/RI feedback will be sent in the uplink direction (which can be sent as periodic or aperiodic CQI as presented in Chapter 3). The eNodeB will then determine type of MIMO transmission to be used. Since in Release 10 the data and reference signal are sent with the same precoder and power level, there is no need for further power offset signalling as used for transmission mode 5 based MU-MIMO in Release 8. Including power offset signalling implicitly in the reference signal also allows more flexible resource allocation for MU-MIMO. With the URS based solution MU-MIMO pairing decision can, in theory, be done independently for each PRB and this allows the eNodeB to achieve both frequency selective scheduling gain and MU-MIMO gain within the same transmission mode. In practice there are some limitations on the pairing due the limited flexibility in allocating URS ports and sequences.

The MU-MIMO operation has been also improved with the better feedback operation and also the dynamic switching between SU-MIMO and MU-MIMO is also enabled with the TM9 in Release 10 as also introduced in Release 9 for TM8. The feedback supports both downlink SU-MIMO and MU-MIMO operation. The different use cases for the SU-MIMO have been identified as follows in 3GPP:

- Closely-spaced cross-polarized antennas.
- Closely-spaced Uniform Linear Arrays (ULA) type antenna installations, which have higher correlation between the antennas compared to the cross-polarized antennas, especially of the cross-polarized antenna have wider spacing.
- Widely spaced cross-polarized arrays, which have little or no correlation between the antennas.

The lack of correlation makes the scenario more suited for a SU-MIMO operation with higher rank while more correlation is then suggesting towards MU-MIMO type of operation as presented in the Section 6.5 performance studies.

During the Release 10, a large number of investigations related to code book design were carried out. In the end it was agreed that only the case of 8tx code book, which is not supported in earlier releases, needed to be addressed in terms of changes to the standard. No changes for the 2tx and 4tx codebooks were made and the only change to the overall UE feedback solution from Release 8 for 2tx and 4tx was the use of CSI-RS for channel measurements in transmission mode 9.

The 8tx codebook feedback was designed based on the 'double codebook' principle in such a way that it can be divided into short-term feedback and long-term feedback. The short-term feedback is more relevant for the SU-MIMO operation while the long-term feedback is then having more static elements, reflecting the long term parameters such as angular spread of the transmission. 3GPP did the following in Release 10: MU-MIMO feedback is optimized for providing feedback on the UE separation in spatial domain, reflecting the long-term channel wideband correlation properties which do not change that rapidly between individual CSI reports.

The double code book structure can be considered to be operating as follows:

- The precoder W for a particular sub-band is consisting of two parts with W_1 targeting for the mentioned long term wideband channel properties while the W_2 corresponds for the short term frequency selective co-phasing CSI.
- The feedback rate can be adjusted so that the W_1 part is updated less frequency as it is expected to change also less often. Also sub-band specific feedback for W_1 is not needed while it is very important for W_2.

The double code book (also sometimes referred as two-stage codebook) structure has been added to the 8×8 case. The double codebook principle is illustrated in Figure 6.7, with the UE potentially providing more frequency feedback on the short term PMI and CQI information.

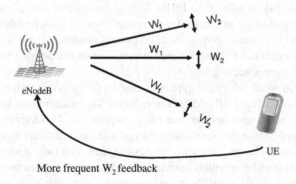

Figure 6.7 Double codebook structure with TM9.

With the two- and four-antenna transmission mode the existing Release 8 code books remain valid as they provide quite a good performance. Further, the CSI reporting has been rather similar with the Release 8 principles only accommodating the two-stage feedback structure. Both the wideband and sub-band periodic feedbacks have been extended to also cover the eight antenna case.

The UE categories can now indicate MIMO capability per frequency band, which is especially important when considering the different nature of the frequency bands. With the bands below 1 GHz the benefits from the further diversity antennas are fewer and implementing extra antennas with decent performance more challenging than at the bands around 2 GHz or above. Thus the UE may indicate for example support of two antennas with 800 MHz and four antennas at 2.6 GHz if that helps the implementation. The UE categories are addressed in further detail in Chapter 12.

6.5 Performance

In this section the benefits from the downlink MIMO operation aggregation are investigated, both theoretically and also based on the simulations. The following areas are improved:

- Higher peak data rate,
- Improved capacity.

The higher peak data is obviously enabled with the addition of the number of parallel transmission streams numbering up to eight. With a 20 MHz carrier, thus the peak data rate would be increase from the 150 Mbps with 2×2 MIMO, all the way to the 600 Mbps with 8×8 MIMO, while the data rate with the 4×4 MIMO is 300 Mbps. Use of 8×8 MIMO would allow the theoretical data rate of up to 3 Gbps with the use of up to five carriers.

Two downlink MIMO features in LTE-Advanced need to be highlighted as providing improved capacity:

- Enhanced support for MU-MIMO operation;
- UE feedback for eight transmitter antenna.

The enhanced MU-MIMO support is obtained by using UE specific reference signals. This allows for full flexibility in allocating PRBs amongst users and also allows for optimized precoding which potentially can reduce interference amongst co-scheduled users. The performance improvement is most obvious with co-polar antennas at the eNodeB where a significant gain over the optimal Release 8 transmission scheme can be obtained already in case of four transmitter antenna, see Figure 6.8.

For cross-polar antennas, which are typically used in the field due to the reduced size, the performance gain from adding MU-MIMO is reduced. The main reason is that MU-MIMO provides gains when multi-user interference can be suppressed. This happens best in the case of co-polar antennas because the correlation amongst antennas remains high, and that allows shaping of robust narrow beams targeting the desired user with only limited interference to other users. With cross-polar antennas correlation is decreased and beam-shaping capability is reduced. In theory it is still possible to suppress multi-user interference but it is less efficient and more sensitive to inaccuracies in the channel state information. In Figure 6.9 the

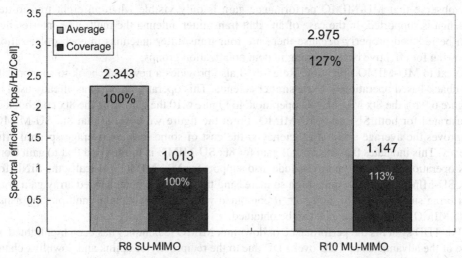

Figure 6.8 Performance enhancement from Release 10 four antenna MU-MIMO, co-polar antennas assumed at the eNodeB.

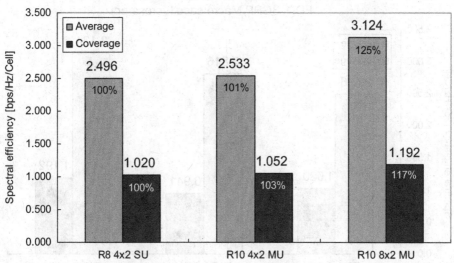

Figure 6.9 Performance enhancement from Release 10 MU-MIMO, cross-polar antennas assumed at the eNodeB.

performance of Release 10 MU-MIMO with cross-polar antennas is shown. From the figure we observe that MU-MIMO performance gain is only visible when an eight transmitter antenna is supported. In the case of an eight transmitter antenna the multi-user interference can be reduced properly because there are four transmitter antennas in each polarization allowing for effective beam-shaping in both polarization groups.

Next to MU-MIMO operation, Release 10 also provides a new code book to support UE feedback-based operation of 8 transmitter antenna. This operation mode was already used in Figure 6.9 for the 8tx MU-MIMO operation. In Figure 6.10 the gain from the 8tx codebook is illustrated for both SU- and MU-MIMO. From the figure we observe that 8tx SU-MIMO improves the average spectral efficiency at the cost of some loss in coverage spectral efficiency. This indicates that the overall gain for 8tx SU-MIMO is limited and that to gain from 8tx operation it is important to include also support for MU-MIMO. Basically the SINR for 4tx SU-MIMO is already quite high so increasing that further with enhanced array gain does not bring substantial gain; however, if combined with additional spatial multiplexing using MU-MIMO, performance gain can be obtained.

For TDD systems the performance of downlink MIMO techniques has been highlighted as one of the advantages of TDD over FDD due to the reciprocity of uplink and downlink channels (see [1]). With Release 10 we have now multiple options for eNodeB to acquire spatial channel state information for 8tx TD-LTE system. One is based TDD channel reciprocity and the uplink sounding reference signal, another is based on UE feedback and the newly designed 8tx codebook. For the uplink sounding based solution it is important to know what exact type of sounding is considered. Especially whether UE supports transmitting sounding from both antennas seems to be important for the final performance. In Figure 6.11 we show a performance comparison for three different operation modes, the first is based on PMI

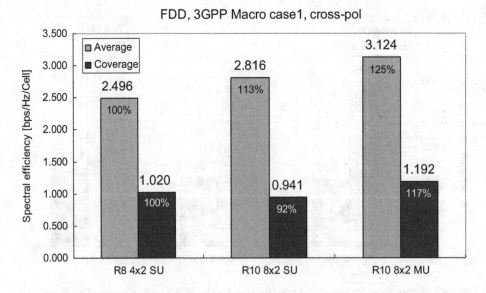

Figure 6.10 Release 10 performance gain from the 8tx codebook.

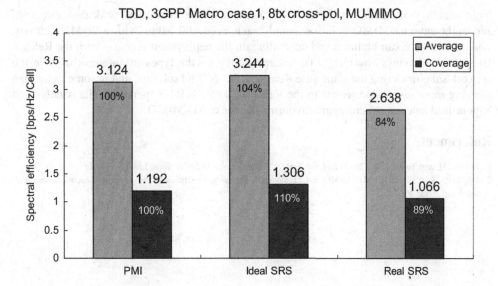

Figure 6.11 Performance comparison between different UE feedback methods for TD-LTE supported by Release 10.

using aperiodic PUSCH feedback mode 3–2, two others use uplink sounding. For uplink sounding we consider one option supporting antenna switching (ideal) and one not supporting (real). From the results in Figure 6.11 we can see that with the ideal sounding TDD-style downlink MIMO gives slightly better performance than PMI-based. However, for the case-realistic case where UE antenna switching is not enabled we have a non-negligible loss compared to the PMI based option. In conclusion, the TDD channel reciprocity does, under the right circumstances, provide some benefits over PMI-based feedback but PMI-based solution offers a feasible alternative which does not require UE to support transmitter antenna switching and thus also lowers the uplink overhead from sounding.

Overall Release 10 downlink MIMO provides enhanced performance for a number of configurations. Benefits are most apparent in case of eight transmitter antennas which were not well supported in earlier releases, but for four co-polar antennas the new flexible MU-MIMO operation also improves performance.

The situation with the MU-MIMO is changed, however, when moving to the non-correlated antennas when not that much gain can be achieved, as has been studied already for the Release 8 MU-MIMO and reported in the results in [2].

6.6   Conclusions

In this chapter, we have covered the LTE downlink MIMO enhancement principles as defined in 3GPP Release 10 specifications as well as the resulting performance. 3GPP has made further investigations in Release 11 to see whether there would be some additional improvements giving enough capacity gain to justify the additional complexity for supporting them, as discussed in Chapter 12. Downlink MIMO as a feature is already widely adopted in the field and also the experiences of the Release 8 performance are good showing large usability of MIMO in the practical networks. The theoretical peak data rate with and eight antenna MIMO solution could reach even 600 Mbps with a 20 MHz carrier. The performance can be improved depending on the deployment scenario with the Release 10 solutions having, however, a large dependency on the types of antennas used in the network side, favouring the same case Release 8 SU-MIMO solution, and in some cases then showing more favourable results to the Release 10 MU-MIMO operation. The widely-used X-polarized antenna structures are favouring the use of SU-MIMO.

References

1. Holma, H. and Toskala, A. (2011) *LTE for UMTS*, 2nd edn, John Wiley & Sons, Ltd, Chichester.
2. Duplicity, J. *et al.* (2012) MU-MIMO in LTE systems. *Euraship Journal on Wireless Communications and Networking*, 2011: 496763.

7

Uplink MIMO

Timo Lunttila, Kari Hooli, YuYu Yan and Antti Toskala

7.1 Introduction

This chapter presents the LTE uplink MIMO included in Release 10 LTE specifications to enable multi-stream multi-antenna uplink transmission as part of LTE-Advanced efforts in meeting the IMT-Advanced and LTE-Advanced requirements. This chapter contains firstly an overview of the uplink MIMO development from Release 8 virtual MIMO to Release 10 MIMO solutions. Then the changes in the protocol side are covered and the impacts to the physical layer are presented, while also addressing some of the design choices with uplink multiple access development done in connection with the LTE-Advanced uplink MIMO studies. This chapter finally concludes with the uplink MIMO performance investigation.

7.2 Uplink MIMO Enhancements Overview

In contrast to the downlink MIMO operation in LTE Release 8, there is no multi-antenna uplink as part of the Release 8 or 9 LTE specifications. In the early days of LTE studies uplink MIMO was also considered but then it was decided to focus the Release 8 LTE specification work in the downlink-only MIMO solutions. In the uplink direction there is already the possibility to perform so called virtual MIMO or uplink Multi-User MIMO (MU-MIMO) in Release 8 anyway by allocating two users the same set of physical resource blocks, while those users are transmitting using mutually orthogonal reference signal (RS) sequences. Additionally, antenna selection transmit diversity was included in Release 8 as an optional feature. Release 10 now provides a lot of extra functionality both to add more flexibility to the earlier defined MU-MIMO operation as well as to introduce the Single User MIMO (SU-MIMO) in the uplink direction as well. The following key new areas are introduced on top of the Release 8 MIMO capabilities as shown in Figure 7.1.

LTE-Advanced: 3GPP Solution for IMT-Advanced, First Edition. Edited by Harri Holma and Antti Toskala.
© 2012 John Wiley & Sons, Ltd. Published 2012 by John Wiley & Sons, Ltd.

LTE Release 8	**LTE Release 9**	**LTE Release 10**
Virtual MIMO (MU-MIMO) support with orthogonal reference symbols allocated for the UE	No uplink MIMO enhancements	Support for SU-MIMO with up to 4 layers (4x4).
Optional antenna selection transmit diversity		More flexible MU-MIMO with flexible reference symbol and PUSCH allocation.

Figure 7.1   LTE uplink MIMO developments in different releases.

- Introduction of the two- and four-antenna uplink SU-MIMO transmission, with the overall principle rather similar to the downlink MIMO operation, with the two-antenna MIMO operation illustrated in Figure 7.2.
- More flexible reference signal sequence allocation resulting to more flexible pairing for the users for virtual MIMO (MU-MIMO).
- Necessary downlink signalling to facilitate eNodeB based precoder selection for the UE transmission.

The solutions introduced enable an increase in the uplink peak data rate per carrier, and add more flexibility for the uplink MU-MIMO operation with single transmit antenna UEs while remaining fully backwards-compatible with the Release 8 and 9 terminals.

7.3   Protocol Impacts from Uplink MIMO

Similar to the downlink MIMO, the uplink MIMO operation mostly impacts the physical layer. However, uplink MIMO impacts also on the other layers. The uplink MAC layer scheduler in the eNodeB needs to consider the UEs MIMO capability when making scheduling decisions. Similarly, if there is a desire to use the Virtual MIMO (MU-MIMO) feature in the uplink, the scheduler needs to consider which users can be paired together. The requirements for the pairing are now relaxed compared to Release 8 thanks to the more flexible reference signal allocation, enabling different users to overlap only partly in frequency as is

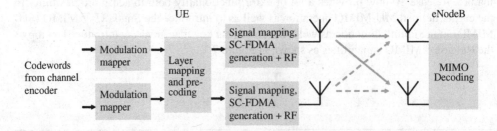

Figure 7.2   Overall MIMO principle in the uplink.

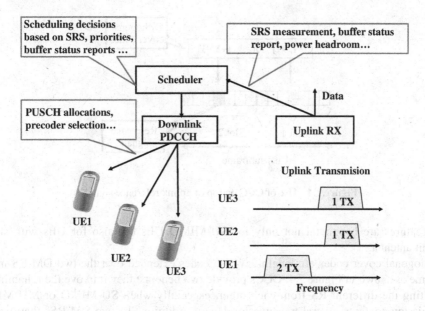

Figure 7.3 Uplink MIMO operation with eNodeB scheduler.

addressed in Section 7.4 in more detail. MAC layer operation in connection with uplink MIMO is illustrated in Figure 7.3.

The RRC signalling is used to configure the parameters related to uplink MIMO. In Release 10, multi-antenna transmission is configured separately for PUSCH, PUCCH and SRS. RRC signalling is used also for enabling the uplink reference signals enhancements as covered in Section 7.4 as well as for the configuration of both periodic and aperiodic SRS for the UE.

7.4   Physical Layer Impacts from Uplink MIMO

Moving from a single antenna transmission to two- or four-transmitter antennas required some partly new structures in the physical layer. It also required consideration for the resulting impact to the signal waveform properties. Also, the reference signal structure needed modification to cover the case of multiple transmit antennas per UE as well as to enable better use of the MU-MIMO functionality.

7.4.1   Uplink Reference Signal Structure

In the uplink direction there are two additional solutions introduced to the reference signal structure:

- Orthogonal Cover Codes (OCC) to complement the use of cyclic shifts of the Demodulation Reference Signal (DM RS) sequence;
- Aperiodic Sounding Reference Signal (SRS) to enhance the use of available SRS resources.

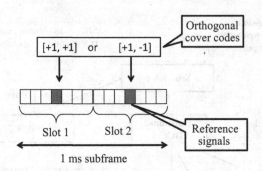

Figure 7.4 Use of OCC on top of uplink reference signals.

Both features are beneficial not only for SU-MIMO UEs but also for UEs with single transmit antenna.

Orthogonal cover codes are length-2 Walsh codes extended over the two DM RS in the subframe as shown in Figure 7.4. OCCs provide two benefits: they improve the reliability in separating the different RS from each other, especially when SU-MIMO or MU-MIMO transmission contains several transmission layers. Additionally, two DM RSs that overlap only partially in frequency are sufficiently separated. This is not the case when DM RS are separated by cyclic shifts. This feature is especially beneficial in the case of MU-MIMO, as now the PRB allocations of the paired UEs do not need to be fully aligned, just for the sake of sufficient DM RS separation. When OCC is used, the eNodeB needs to signal to UE both cyclic shift and orthogonal cover code. Further, when SU-MIMO transmission has several transmission layers, a separate DM RS is needed for each layer. The eNodeB can indicate eight different sets of cyclic shift and OCC combinations with a 3-bit field included to the uplink-related DCI formats. The cyclic shift and OCC combinations in a set are designed to provide sufficient DM RS separation between transmission layers. UE simply uses a part of the set corresponding to the number of transmission layers. Finally, OCC cannot be used together with sequence group hopping, which can be disabled with UE-specific higher layer signalling in LTE Release 10.

The same precoding as with the uplink data is applied on the DM RS, while the SRS transmission is not precoded. This means that the selection of precoder is based on SRS transmissions, and hence, uplink MIMO operation increases the use of SRS. In Release 8 and 9, SRS is configured by RRC signalling and it is transmitted periodically, regardless of the momentary need for channel sounding. To enhance the use of SRS resources, aperiodic SRS is introduced in Release 10. Aperiodic SRS is configured by RRC signalling, but actual SRS transmission is triggered by an indication on the PDCCH. The SRS transmission can be now more dynamically controlled by the eNodeB. The SRS transmission can be focused to those UEs which have indicated the need for transmission (scheduling request) or which are known to have lot of data in the buffer based on the MAC layer buffer status reports as covered in Chapter 3. Aperiodic SRS request can be configured to DCI formats 0, 1A, 4 and, in the case of TDD, also to DCI formats 2B and 2C. In other words, aperiodic SRS can be triggered by both downlink assignments as well as uplink grants. To allow for more flexible aperiodic SRS use, UE may have multiple aperiodic SRS configurations. In DCI format 4, the aperiodic SRS request is a 2-bit field that is one from three different SRS configurations.

Separate SRS configurations are also associated to 1-bit aperiodic SRS requests in DCI format 0 and in DCI format 1A/2B/2C. Multiple aperiodic SRS configurations can be used to simplify aperiodic SRS multiplexing with other UEs or to support aperiodic sounding for multiple purposes with different sounding bandwidths, number of antenna ports, and so on.

In the case of multiple antenna SRS configuration, SRS transmission from different antennas is simultaneous and SRSs are separated by the different cyclic shifts and transmission combs.

7.4.2 MIMO Transmission for Uplink Data

The uplink SU-MIMO with two or four antennas was introduced to improve the uplink performance in order to meet the LTE-Advanced requirements. The solution adopted is based on the wideband (i.e. non-frequency selective) closed-loop precoding. The wideband solution was chosen as it was considered that the frequency selective precoder feedback not giving additional benefits, and only increased the overheads and complexity further. Also, the open loop solution was considered but there were no real benefits to adding multiple options in the specifications as the closed loop approach performed better anyway as shown in [1]. The closed loop uplink MIMO operation is based on the eNodeB making estimates based on the received SRS (which was sent without precoding) and then signalling to the UE the precoder to be used for the uplink transmission on PUSCH. The signalling is part of the uplink grant signalling on the PDCCH as shown in Figure 7.5.

In LTE Release 10, there are two transmission modes defined for PUSCH. Transmission mode 1 (TM1) uses single antenna port (SAP) transmissions, which can be scheduled with DCI format 0 uplink grant. It can be used by both SU-MIMO UEs as well as UEs having only a single transmit antenna. Naturally, TM1 is also the default transmission mode that is used for example during random access procedure. In single antenna port transmission, the PUSCH signal received at eNodeB should appear as if transmitted from single antenna. The

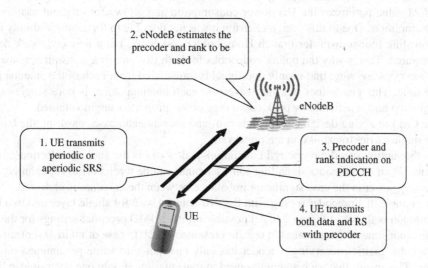

Figure 7.5 Closed loop uplink MIMO operation.

actual implementation is left for UE vendors to decide; a UE may select only one antenna for transmissions, or transmit the same data from all antennas with a fixed precoding.

Transmission mode 2 (TM2) was introduced for uplink SU-MIMO transmissions. In TM2, eNodeB can dynamically select between two transmission schemes when scheduling uplink for UE:

- Single antenna port transmission. SAP transmission can be used when an eNodeB does not have up-to-date channel information to decide on precoding and transmission rank. It is used as a fall-back transmission scheme as well as with semi-persistent scheduling.
- Closed-loop spatial multiplexing transmission. Here, a UE may transmit single transport precoded over available antennas or spatially multiplex two transport blocks. This transmission scheme is considered further in following.

In the closed-loop spatial multiplexing, the UE uses at most two transport blocks per subframe, with the respective feedback provided on the PHICH in the downlink indicating whether each one of them was received correctly. The uplink MIMO transmission are scheduled using the new DCI format 4 uplink grant which contains the following new information for uplink MIMO support:

- Modulation and coding scheme for each transport block.
- Precoding information including the number of spatial layers as well as the selected precoder.
- New data indicator for each of the transport blocks separately to indicate if retransmission is expected or not.

The codebook design for the uplink reflects the key differences between uplink and downlink multi-antenna operation. Similarly, as in any uplink considerations, in the case of uplink MIMO the low Cubic Metric (CM) of transmitted waveform also plays a significant role. Low CM value optimizes the UE power consumption and allows for efficient usage of all power amplifiers. Due to this design criterion it was not possible to for example simply reuse the downlink four-transmitter branch Householder codebook, but a new codebook design was required. This is why the uplink codebooks for both two and four transmit antennas are *cubic metric preserving*; that is only one signal is transmitted from each of the antenna ports at the time. This guarantees that the waveform each antenna sends is pure single-carrier transmission and beneficial low peak-to-average power properties are maintained.

The CM preserving design covers both two- and four-antenna cases ensuring the benefits of the single carrier transmission are preserved.

Another uplink specific property in the uplink codebooks is the antenna selection/turn-off elements. With the precoder signalling some antennas can be turned off to for example save UE battery power in the case significant imbalance between the antennas is observed.

An example of a precoder setting with the two-antenna case for single layer and two layer transmission is shown in Figure 7.6. All possible uplink MIMO precoder settings for the case with the four-transmitter antennas in use are presented in [2]. In case of multi-layer transmission, a cubic metric preserving precoder has only one non-zero value per antenna on each column. This means that each antenna is used in transmission of only one layer. It also limits available precoders to only one possible precoder for rank 2 transmission in a 2tx case, as

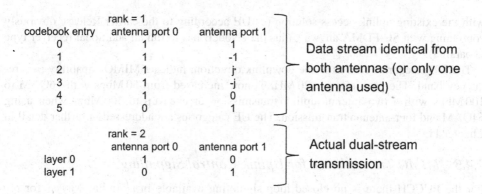

Figure 7.6 Uplink precoder settings for a two transmit antenna case.

well as for the rank 4 transmission in 4tx case. Of course, the standardized codebooks contain more alternatives from which the precoder can be selected and then signalled to UE when the rank used is less than the number of antennas in use. As in the downlink direction, the precoding will help to maximize the signal-to-noise ratio (SNR) at the receiver when suitable precoder is selected according to the channel properties.

On the transmitter side, the data from channel encoder is mapped to layers used, with the example of the two-antenna uplink transmitter shown in Figure 7.7. Each transport block is mapped on one spatial layer in case of two antennas while in case of four antennas, the transport block is mapped either on one or two spatial layers, depending on the transmission rank. Each layer is used to carry data for only one transport block.

As part of the LTE-Advanced studies it was also considered whether OFDMA in the uplink would be better suited together with the multi-antenna transmission. The studies however, indicated, as also shown in [3], that the same performance can be achieved with the SC-FDMA as well. Thus adopting OFDMA uplink multiple access option was not justified, rather use of that in the uplink would have only lead to the increase in the cubic metric as presented also in [4]. Also it would have required an additional transmitter type in the UE and an additional receiver type in the eNodeB as the transmission would not be compatible

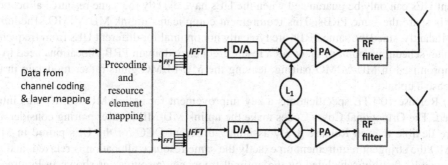

Figure 7.7 Uplink MIMO transmitter.

with the existing uplink access solution (as UE according to the earlier Release obviously continuing with SC-FDMA anyway, thus the eNodeB has to run an additional receiver type in parallel).

The UE categories can, as in the downlink direction, indicate MIMO capability per frequency band. The peak rate per 20 MHz is now increased from 50 Mbps with 16QAM to 100 Mbps with a two-antenna uplink transmission or even up to 300 Mbps when using 64QAM and four-antenna transmission. The UE categories are addressed in further detail in Chapter 11.

7.4.3 MIMO Transmission for Uplink Control Signalling

For the PUCCH there is no closed loop signalling available in a similar way as for the PUSCH. Since the UE needs to send the PUCCH without any downlink feedback, a closed loop solution could not be applied. Thus the design choice was to use transmit diversity to benefit from the situation when the terminal is able to use more than one antenna for the PUCCH transmission. The principle is based on the Space-Orthogonal Resource Transmit Diversity (SORTD) which basically means that orthogonal PUCCH resources are allocated for both of the antenna ports used (two antenna ports are used regardless whether two or four antennas are available) and then the same data are transmitted from both antenna ports. The SORTD is applicable for the PUCCH formats 1/1a/1b, 2/2a/2b and 3. SORTD allows for full transmit diversity and hence provides improved uplink signalling performance.

When uplink control information is multiplexed on PUSCH with multiple transmission layers, two different approaches are taken depending on the type of control information. HARQ acknowledgements for PDSCH and downlink rank indicators are replicated and transmitted on all transmission layers. The purpose is to improve reliability for this critical control information. The position of control symbols is also time aligned over the layers, facilitating for the use of advanced receiver for HARQ-ACK and RI detection. On the other hand, the typically larger downlink channel quality indicator and precoding matrix indicator are transmitted only on a layer (or layers) corresponding to single transport block.

7.4.4 Multi-User MIMO Transmission in the Uplink

LTE Release 8 Multi-User MIMO operation is completely transparent to the UE: the eNodeB may simply schedule two UEs on the same resources without the UEs being aware of it. However, due to the uplink reference signal properties orthogonality between the RS of different UEs can only be guaranteed when the UEs have exactly the same resource allocation, that is share the same PRBs. This requirement complicates uplink MU-MIMO scheduling significantly, since the same PRBs are typically not optimal for different UEs from frequency domain scheduling point of view. As a result, frequency domain PRB allocations need to be compromised in MU-MIMO pairing, leaving the MU-MIMO gains rather moderate in the Release 8 uplink.

In Release 10 LTE specifications a key improvement for uplink MU-MIMO was introduced. The Orthogonal Cover Codes make the uplink MU-MIMO user pairing considerably more flexible than in Release 8. By assigning different OCC for the UEs paired in MU-MIMO the stringent requirement on exactly the same resource allocation is relaxed, and the UEs can be flexibly scheduled on the optimal frequency resources as shown in Figure 7.8 while the eNodeB can still separate the transmission from different users overlapping in the

Release 8 MU-MIMO resource allocation

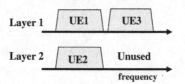

Release 10 MU-MIMO resource allocation

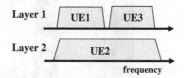

Figure 7.8   Uplink MU-MIMO resource allocation in Release 8 and 10.

frequency domain. This in turn helps in increasing the probability for finding UEs for
MU-MIMO operation without need to make compromise on PRB allocations. This, especially with multi-cluster scheduling, avoids the need to leave empty resources and hence
also improves the uplink average cell throughput by up to 15% [5].

7.5   Performance

In this section the benefits from the uplink MIMO operation are investigated, both theoretically and also based on the simulations. The following areas are improved:

- The uplink peak data rate achieved is increased with 16QAM modulation from 50 Mbps
  per 20 MHz to 100 Mbps and with 64QAM respectively from 75 Mbps to 150 Mbps in
  case of two antennas and further up to 300 Mbps with 4 stream MIMO transmission.
- The uplink capacity can be increased up to 40% compared to single antenna transmission,
  depending of the number of antennas used and on the correlation properties between
  the antennas.
- Further the MU-MIMO performance is improved as the flexibility of the pairing of users
  was improved.

Uplink throughput improvements from MU-MIMO and SU-MIMO are illustrated in
Figure 7.9. In the results, the throughputs with and without MIMO operation are compared
for the same number of receive antennas at the eNodeB. In the figure, gains from SU-MIMO
as well as from MU-MIMO with Release 10 DM RS enhancements are shown. It can be
noted that both SU-MIMO and MU-MIMO provide comparable gains on the cell capacity.
Additionally, SU-MIMO precoding enhances considerably cell edge performance. Throughput gains from MIMO depend on a number of factors. For example, inter-cell interference
rejection combining at eNodeB, which was not considered in here, typically reduces achievable MIMO gains. On other hand, MIMO gains for cell capacity are increased if the portion of
UEs supporting 64QAM is decreased.

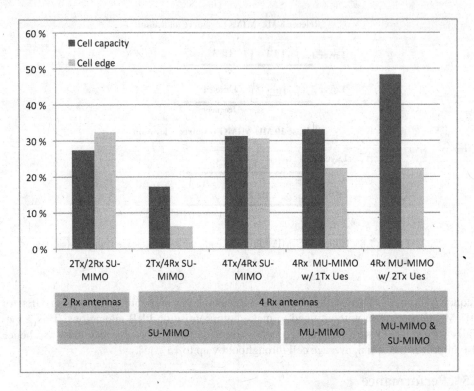

Figure 7.9 SU-MIMO and MU-MIMO throughput gains.

The MU-MIMO performance thus improves as users are easier to schedule on overlapping resources when the PRBs do not need to be exactly the same, thanks to the use of OCCs with the uplink RS. The performance is investigated in Figure 7.10 for a 10 MHz case with a four-antenna MMSE receiver at eNodeB. Such a configuration could lead to up to a 40% improvement over the single user case without MU-MIMO. With the low number of users there are not that many possibilities how to pair different users, thus the gain will increase (and then saturate) up to 10 users. At that point the scheduler has enough possibilities to choose for uplink scheduling.

There is further potential to improve the performance by combining the MU-MIMO with uplink multi-cluster scheduling. The resulting scheduler flexibility enables getting even more out of the MU-MIMO operation adding further on top of the MU-MIMO performance.

7.6 Conclusions

In this chapter, we have covered the LTE uplink MIMO principles as defined in 3GPP Release 10 specifications as well as the resulting performance. The use of uplink SU-MIMO allows improving the peak data rate and performance but the price to pay is the need to use of two or more uplink transmitter RF chains, including use of more than one power amplifier and UE transmit antennas. The implementation is made a bit easier by using such precode design which retains the signal waveform properties (cubic metric) to avoid additional power

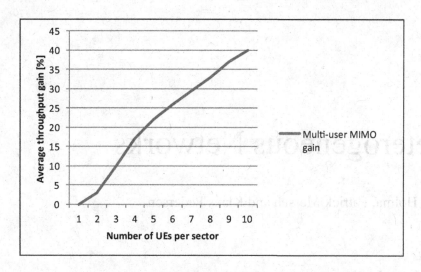

Figure 7.10 Uplink MU-MIMO gain as the function of users per sector.

reduction. In addition to the capacity the theoretical uplink peak data rate per 20 MHz carrier is increased up to 300 Mbps. For Release 11 there is no further work ongoing on the uplink MIMO enhancements but as part of the work on the Co-ordinated MultiPoint (CoMP) work item, the uplink aspects of the multi-site reception (and possible related feedback) are being addressed for Release 11 as covered in Chapter 13.

References

1. Lunttila, T., Kiiski, M., Hooli, K. *et al.* (2009) Multi-antenna techniques for LTE-advanced, Proceedings of WPMC 2009, Sendai, Japan.
2. 3GPP Technical Specification, TS 36.211 (December 2011) Physical Channels and Modulation, v 10.4.0.
3. Berardinelli, G., Navarro Manchón, C., Deneire, L. *et al* (2009) Turbo Receivers for Single User MIMO LTE-A Uplink, IEEE.
4. Holma, H. and Toskala, A. (2011) *LTE for UMTS*, 2nd edn, John Wiley & Sons, Ltd, Chichester.
5. 3GPP Tdoc R1-094651 (November 2009) Performance of uplink MU-MIMO with enhanced demodulation reference signal structure, Nokia Siemens Networks, Nokia.

8

Heterogeneous Networks

Harri Holma, Patrick Marsch and Klaus Pedersen

8.1 Introduction

The traffic volumes in mobile broadband networks have increased considerably during recent years and this growth is expected to continue. While network capacity can be increased through transmission schemes achieving higher spectral efficiency or by using more spectrum, as addressed in previous chapters, the most straightforward approach is to deploy more base stations. With more base stations, radio resources can be reused more often per area, effectively increasing the spectral efficiency per area. Naturally, a mobile network operator will gradually deploy additional base stations where the traffic demand is highest, and choose antenna configurations and transmit powers in order to optimally complement the existing cell infrastructure at minimized cost. This typically leads to network configurations with a co-existence of large base stations and new smaller base stations. In addition, the evolution of networks will lead to a co-existence of different Radio Access Technologies (RATs), such as GSM, WCDMA/HSPA and LTE/LTE-Advanced. Constellations of multiple kinds of base station sizes (so-called *layers*) and multiple RATs are often referred to as heterogeneous networks (HetNets), which are to be addressed in this chapter.

What makes small cells particularly attractive is that their cost of hardware is lower than that of large cells, which are typically referred to as macro cells, and technology evolution further pronounces this trend. The reason is that digital signal processing is evolving faster than RF technology, and the extent of RF circuitry is relatively smaller for small base stations than for macro base stations. Further, the transmit power of small base stations is much lower than that of macro base stations, as we will see later, and consequently the overhead required for power amplification and cooling is considerably reduced. Overall, small base stations can be built at a lower cost, have lower power consumption and smaller form factors, which makes their installation and site acquisition easier.

LTE-Advanced: 3GPP Solution for IMT-Advanced, First Edition. Edited by Harri Holma and Antti Toskala.
© 2012 John Wiley & Sons, Ltd. Published 2012 by John Wiley & Sons, Ltd.

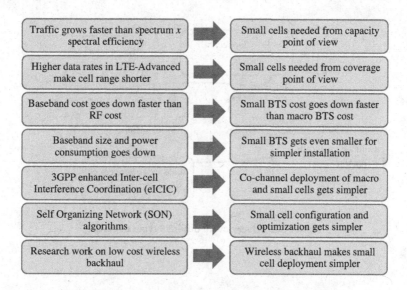

Figure 8.1 Drivers and enablers for small cell deployments.

Clearly, one challenge connected to heterogeneous networks with more and diverse base stations is that these require more sophisticated network configuration and optimization, but we will see in this chapter that LTE-Advanced can help also in this area through the availability of Self-Organizing Network (SON) functionality and enhanced Inter-Cell Interference Coordination (eICIC) algorithms.

Figure 8.1 illustrates the drivers and enablers for small cell deployments. Small cells have been discussed in the research field for many years, but only now we are at the beginning of also seeing small cell deployments in a significantly larger scale in mobile networks.

This chapter is organized as follows. First, Section 8.2 describes the base station classes as defined by 3GPP. Section 8.3 then deals with traffic steering and mobility management in the context of heterogeneous networks; that is, the challenge of how to dynamically assign devices and services to the right layer or RAT. Section 8.4 presents various approaches on how to tackle intra-frequency interference, which is one of the main issues connected to heterogeneous deployments. Section 8.5 then goes into more detail for particular HetNet scenarios, such as outdoor small cells or indoor femto cells, and presents performance results connected to the mobility and interference management schemes discussed before. The chapter is finally summarized in Section 8.6, and references are provided in the final section.

8.2 Base Station Classes

Most of the base stations so far have been wide area macro cells that provide coverage for several square kilometers by using high mounted antennas and high power transmitters. Outdoor small cells are often called micro cells and provide coverage for a few hundred meters by using lower antenna installations. Indoor cells can be located in public premises, like shopping malls or office buildings, or at homes. The public indoor solutions have traditionally used indoor antennas and RF cables connected to a macro base station that is located in

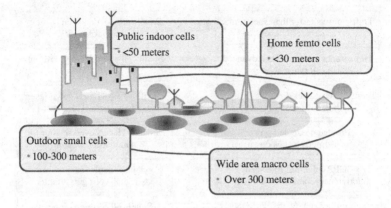

Figure 8.2   Overview of different types of cells.

the same building, a solution called a Distributed Antenna System (DAS). The benefit of a DAS is that upgrading new radios is relatively simple – in the best case, one just has to plug in a new base station to the cable system without any upgrades to the cables or antennas. The drawback of DAS is that adding more antennas may only increase the coverage and provide a more homogeneous quality of service over the coverage area, but it cannot lead to a substantially increased capacity, as the same physically layer transmission resources are simply shared by a larger coverage area. If capacity is to be increased, this has to be done via cell splitting, which is complex and most likely requires major changes to the cabling.

Another indoor solution is a home base station, also known as a femto cell. Femto cells have been mainly deployed for improving the coverage and capacity in homes and in small offices. An overview of different types of cells is shown in Figure 8.2.

3GPP has defined RF requirements separately for wide area, medium range, local areas and home base stations. The base station classes are defined based on how close to the antennas the users are able to get. This is measured as Minimum Coupling Loss (MCL). For wide area base stations, one assumes that the MCL is more than 70 dB, which in practice refers to a high mounted antenna installation, like a rooftop, mast or pole. Medium range base stations assume an MCL of more than 53 dB and local area ones are defined to have an MCL of 45 dB, which means that the user can get very close to the antenna. The maximum power level of a wide area base station is not limited in 3GPP specifications. The typical products with 2×2 MIMO support up to 2×80 W power levels. Additionally, the macro base station typically uses sectorization with three to six sectors which can push the total site power up to 1 kW. That of the medium range, local area and home base stations is limited to 5 W, 0.25 W and 0.10 W, respectively. These power levels are the combined power of all transmission branches together. For example, the limit of 0.25 W is equal to 2×0.125 W or 4×0.0625 W in two and four antenna cases. There are also other differences in RF requirements in frequency stability, spurious emissions, sensitivity, dynamic range and blocking requirements. The RF requirements for small base stations are generally more relaxed than for high power base stations, which make it further possible to reduce the cost of RF sections. LTE base station classes are illustrated in Figure 8.3. The medium range class is under preparation in

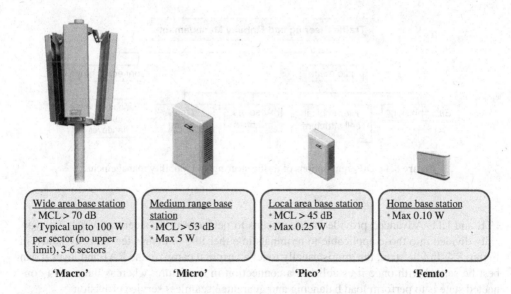

Wide area base station • MCL > 70 dB • Typical up to 100 W per sector (no upper limit), 3-6 sectors	Medium range base station • MCL > 53 dB • Max 5 W	Local area base station • MCL > 45 dB • Max 0.25 W	Home base station • Max 0.10 W
'Macro'	'Micro'	'Pico'	'Femto'

Figure 8.3 3GPP base station classes (MCL = Minimum Coupling Loss between terminal and base station antennas).

Release 11 during the writing of this book, while the other classes are included already in Release 10 [1].

8.3 Traffic Steering and Mobility Management

Traffic steering allows operators to optimize resource utilization, users' service perception and terminal or base station power consumption by directing the traffic to the RAT or layer that provides the best performance regarding any metric of interest. However, this has to be tied closely together with mobility management, which assures robust and optimized mobility performance, for example a reasonable number of handovers and avoidance of radio link failures and ping-pongs.

Especially in the context of heterogeneous networks, traffic steering is a major driver of operational expenses (OPEX) reduction and it enables the limiting or postponement of capital expenses (CAPEX). A common strategy in this respect is to offload as much traffic as possible to femto or WiFi cells, since these typically involve an inexpensive backhaul infrastructure. Furthermore, load balancing between the remaining layers and RATs allows using base station infrastructure more efficiently, hence improving return on investment, and offering a more homogeneous quality of service. Traffic steering needs to take into consideration factors such as:

- signal strength and interference;
- terminal and network capabilities;
- requested services and quality of service (QoS);
- load in different RATs and layers;
- power consumption on the base station and terminal side; and
- terminal speed.

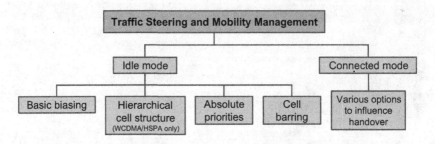

Figure 8.4 Different options of traffic steering and mobility management.

LTE and LTE-Advanced provide various means to perform traffic steering, which are typi-cally divided into those applicable to terminals in either idle or connected state, as shown in Figure 8.4. In idle state, the aim is usually to let a terminal camp on the RAT and layer it can best be served with once it establishes a connection in the future, whereas the aim in con-nected state is to perform load balancing and guarantee seamless service provision.

8.3.1 Traffic Steering and Mobility Management in Idle State

In idle state, that is, where a terminal does not have any radio resource control connection to the radio network, the terminal itself is responsible for measuring its environment and select-ing RAT or frequency layer. However, the network can influence the terminal's decision by providing it with various parameters within so-called system information blocks (SIBs), such as thresholds and priorities involved in the cell reselection procedure. The standardized cell reselection mechanisms in LTE, which can be used for traffic steering purposes, are explained in the list:

- **Basic biasing:** Base stations may broadcast cell selection and ranking parameters to all camped terminals, or provide these during connection release to specific terminals. These parameters contain a cell reselection hysteresis and quality offset which may be adjusted in order to effectively extend or reduce the range within which a cell is likely to be selected by a terminal.
- **Absolute priorities:** The network can also define absolute priorities for different intra- or inter-RAT carriers, again provided to terminals via broadcast or during connection release. The latter option is particularly interesting, as it allows defining absolute priorities for each terminal individually. Higher priority could be used for LTE radio to take full benefit of the new LTE radio network capabilities.
- **Cell barring and black listing:** The network may bar a cell or provide a blacklist of cells on which a terminal may not camp.

One solution to steer more traffic towards smaller cells is to use so-called *range extension*; that is, to virtually increase the size of small cells through basic biasing. This means that a terminal which actually receives a stronger downlink signal from a nearby macro cell than from a small cell connects to the smaller cell nevertheless, as illustrated in Figure 8.5. While this is a very simple and effective means to increase small cell offload, range extension has to

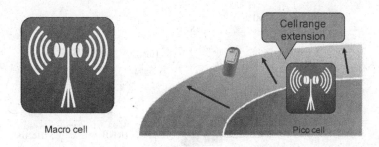

Figure 8.5 Usage of small cell range extension concepts for macro offloading.

be used with care, as too-strong biases can lead to problematic interference constellations, as we will see later.

8.3.2 *Traffic Steering and Mobility Management in the Connected State*

In the connected state, it is the network side which decides upon a handover to be executed between cells, RATs or layers, but this decision is also based on measurements to be performed by the terminal. The network can, however, influence the frequency and scope of these measurements. In LTE and LTE-Advanced, the network can provide a list of carrier frequencies to each terminal, within which the terminal then searches cells autonomously according to given measurement configurations. The network can further influence the scope of these measurements by providing a black cell list containing the physical cell identities (IDs) a terminal shall not measure on certain carriers.

Clearly, a well-tuned and system-wide mobility management requires information exchange between base stations. In LTE and LTE-Advanced, this exchange has to rely on the availability of the X2 interface. In the case of inter-RAT scenarios, the exchange may take place using RAN Information Management (RIM) procedures. In general, one has to assure that all traffic steering and mobility management means used for idle and connected state terminals are well-aligned in order to avoid ping-pong effects or conflicting policies.

8.3.3 *Traffic Steering and Mobility Management with Femto Cells*

Particular challenges arise in the context of femto cells, as these may be deployed in large quantities, and are typically out of the control of an operator: a user may arbitrarily move or turn off a femto base station at any time. Hence, the network has to cope with a continuously changing environment. For a high density of femto cells, the number of possible Physical Cell Identifiers (PCIs) may also start posing a constraint, potentially leading to cell identification ambiguities. Furthermore, femto cells may be constrained to a Closed Subscriber Group (CSG), as mentioned before, reducing cell reselection and handover options.

In LTE and LTE-Advanced, the handling of CSG femto cells is facilitated through the fact that base stations may broadcast PCI ranges connected to CSGs to terminals. Additionally, UEs apply an autonomous search function regardless of which RAT the UEs are is camping on. The autonomous search function determines when and where to search for the allowed CSG cells. Terminals may send these proximity indications to their macro base station,

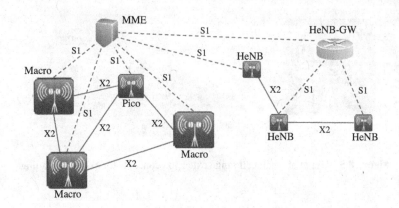

Figure 8.6 Interfaces between different LTE base station nodes for mobility.

stating that they are close to a femto cell to which they could connect. A terminal may for example, trigger such an indication based on historical CSG proximity data it has collected, and which it recognizes through the current carrier activity fingerprint. The exact algorithm, however, is left to the decision of terminal vendors. This option can potentially be used to reduce the number of overall inter-frequency measurements performed by terminals in connected mode.

Furthermore, Release 10 includes mobility enhancements with the introduction of X2 based handovers between Home eNodeBs (HeNB). This is mainly targeted towards Home eNodeB enterprise installations, where Home eNodeBs are deployed with same the CSG ID or configured in Hybrid access mode. Using X2 based handovers between Home eNodeBs reduces the signaling towards the core network, and is in general faster as compared to S1 based mobility procedures. Release 10 does not supported X2 based handover between macro and Home eNodeBs, nor is X2 connectivity via the Home eNodeB gateway (Home eNodeB-GW) supported. The LTE Release 10 HetNet interfaces for mobility purposes are illustrated in Figure 8.6 for different base station nodes.

8.3.4 WiFi Offloading

In practice, almost all smartphones and laptops can be expected to have WLAN (WiFi) connectivity. The data traffic from terminals could be directed to WiFi when a WiFi network connection is available. The offloading of traffic from mobile networks can reduce the congestion in the operator licensed spectrum and improve end user performance. WiFi offloading can happen without any specific network control when the user or the terminal selects a WiFi connection, but there are also various benefits of using so-called network controlled WiFi offloading:

• Automatic switching between cellular and operator *preferred* WiFi network;
• Authentication, authorization and accounting can be provided via WiFi;
• Device management can be offered by the network;
• Integration of WiFi access into a mobile core network.

Many users are unfamiliar with WiFi technology and find it too laborious and complex to make all required selections for getting connected via a WiFi network, thus WiFi networks are not efficiently utilized. The user may have difficulties in selecting the most suitable network when there are multiple WiFi networks in public locations. The network controlled solution makes discovery and selection of WiFi network simpler according to operator and end user preferences. The target WiFi can be a home network, operator hot spot or partner WiFi network. The switching between cellular and WiFi can also be based on the application requirements. For example, when the user launches an application, like a browser, e-mail or Internet radio, the device selects an automatically preconfigured WiFi network, when available, instead of using a cellular network. Therefore, WiFi offloading solutions provide an operator assurance that traffic is offloaded to preferred WiFi networks when available, without having to rely on the user's competence to do so. The automatic selection uses a so-called Access Network Discovery and Selection Function (ANDSF). The purpose of ANDSF is to assist the terminal to discover non-3GPP access networks – that is WiFi – that can be used for communications and to provide the terminal with rules policing the connection to these networks.

The choice between cellular and WiFi networks is similar to the selection of the roaming network when travelling abroad. An automatic selection of the roaming network is based on the configuration that the home operator has provided. However, the user can still override the automatic solution and force the mobile to another roaming network. The same logic applies for selecting Wi-Fi networks – the user may still force the device to use a particular access network, but most users will be happy with the automatic selection.

Authentication, Authorization and Accounting (AAA) can be provided by the operator existing network infrastructure. The authentication can utilize the Subscriber Identity Module (SIM) card and be transparent to end users. Currently, there are various authentication mechanisms in use, for example captive web portal-based login pages requiring username and password, SIM and certificate based authentication and so on. If the user is required to manually input username and password when accessing services via a WiFi network, many users continue to use cellular networks instead of WiFi offloading. Thus, seamless user authentication is an enabler for offloading traffic to a WiFi solution, and it also improves usability.

Device management has been used to provision for example Access Point Name (APN) settings for mobile devices to avoid manual configuration and ensure that important services work flawlessly in mobile networks. WiFi offloading solutions reuse existing device management and extends functional support to provisioning of WiFi network settings, like Service Set Identifier (SSID) or Voice over IP (VoIP) over WiFi settings. Device management can also be used to pre-share keys for encrypted communication.

WiFi access can be integrated to the mobile core network. The integration allows using existing service infrastructure, like charging and policy control, also for the WiFi subscribers. The integration helps also in service continuity when moving between cellular and WiFi. Figure 8.7 illustrates the offloading case with integration to the mobile core network. A Tunnel Termination Gateway (TTG) acts as a gateway towards devices and terminates IP Security (IPsec) connections coming from the terminal to provide data encryption for the WiFi traffic. TTG uses an AAA server for SIM based authentication for the users. After authentication, TTG creates PDP context to selected GGSN. This enables GGSN to reuse existing packet core service infrastructure for WiFi.

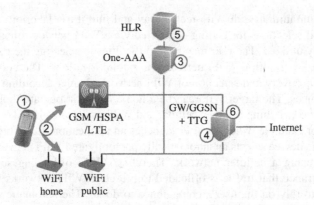

① = Device management configures WiFi settings and network selection rules to the terminal
② = Terminal switches automatically to preferred WiFi network
③ = Authentication for WiFi controlled access
④ = Terminal uses assigned IP address to start interworking procedures towards TTG
⑤ = User authentication with AAA and HLR
⑥ = GW/GGSN allocates IP address for PDP context

Figure 8.7 WiFi offloading solution with integration to mobile core network (TTG = Tunnel Termination Gateway, AAA = Authentication, Authorization and Accounting).

Note finally that 3GPP has not specified any radio level interworking or handovers between 3GPP networks and WiFi networks. Instead, the interworking happens solely on the core network level.

8.4   Interference Management

It is well-known that interference poses the main limitation in today's mobile communication systems, in particular for urban deployments where base station distances are typically on the order of less than a kilometer, and where all radio resources are reused in all cells. While this holds for both homogeneous and heterogeneous deployments, the fact that large and small cells in a HetNet use very different transmit powers raises particularly challenging interference issues, which will be discussed in this section.

Consider a simplified multi-layer setup with a large and small base station as depicted in Figure 8.8, and assume for now that both cells reuse all available radio resources (so-called *co-channel deployment*). The small base station will use a significantly lower downlink transmit power than the macro base station. As terminals by default connect to the cell from which they receive the strongest downlink signal, this means that the cell border will be much closer to the small base station than the macro (indicated through a bold line). If range extension is used, the cell border will be shifted towards the larger base station (thin line). The point where the uplink signals originating from a terminal are received equally strongly at both base stations, however, will be more or less in the middle between these (indicated through a dashed line and denoted as *virtual uplink cell border*). The fact that we have these

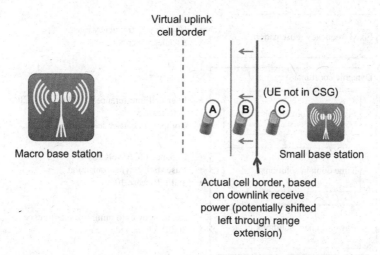

Figure 8.8   The most prominent interference problems in multi-layer networks.

different cell borders in uplink and downlink leads to the following particular interference constellations:

Constellation A: A terminal in location A will be connected to the large cell, but its uplink signals are received more strongly by the small base station than by its counterpart. As it is fairly far away from its assigned base station, it will typically use a high uplink transmit power to overcome path loss, and hence it will create *strong uplink interference to the small base station*. But, it will typically only be scheduled in a fraction of the total bandwidth, and therefore not cause uplink interference problems on all Physical Resource Blocks (PRBs).

Constellation B: A terminal in location B will, in the case of a range extension, be connected to the small cell. However, it actually receives a stronger downlink signal from the macro base station than its assigned base station; hence both data transmission and control channels are subject to *strong downlink interference*. Especially the interference on the control channels is a severe issue which can pose a limitation to the usage of range extension, unless interference coordination solutions are used that we will cover later.

Constellation C: An extreme case of interference arises if a terminal is in location C, but cannot connect to the small cell, as this is limited to a closed subscriber group the terminal does not belong to. Then, the terminal creates very strong uplink interference to the small cell, and is subject to very strong downlink interference from the same. This is typically referred to as a *macro coverage hole*.

Finally, an interference constellation not explicitly shown in the figure, but still of major importance, may arise when multiple small cells reside within or adjacent to a macro cell. In this case, the sum of many uplink signals from terminals connected to small cells may pose detrimental uplink interference to the macro cell.

Figure 8.8 also illustrates the impact of range extension on interference issues. While a stronger range extension increases the probability of interference constellation B (i.e., more

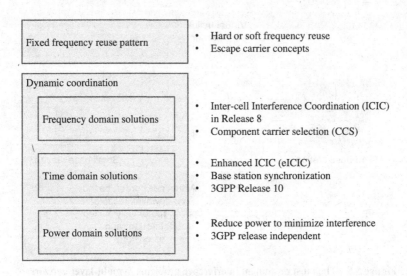

Figure 8.9 Interference management solutions in LTE-Advanced.

severe downlink interference issues), it reduces the probability of constellation A (i.e., alleviates uplink interference problems), as the actual cell border is moved closer to the virtual uplink cell border. It has to be noted that interference does not only impact data transmissions, where link adaptation may potentially be used to mitigate such problems, but it also affects control channels, and may hence lead to the fact that a terminal loses its entire connectivity to a base station.

The question is now how the aforementioned interference constellations can be avoided or at least alleviated. In principle, one has the following options:

- Avoid inter-layer interference by deviating from a full resource reuse amongst cells and introduce radio resources that are dedicated to certain layers. This can be done in:
  - time and frequency, and in
  - a static or dynamic way.
- Apply specific power control rules to alleviate inter-layer interference.

Potential solutions in LTE and LTE-Advanced are summarized in Figure 8.9, and will be elaborated upon in the following sections.

8.4.1 Static Interference Avoidance through Frequency Reuse Patterns

If an operator has multiple carriers or frequency bands available, the most straight-forward way to address the interference problems is to deviate from a *co-channel deployment* and introduce specific frequency reuse patterns. The simplest of these is a so-called *hard frequency reuse*, where macro cells and small cells simply use separate frequency bands, effectively avoiding all of the interference constellations stated before. However, only a few operators have the luxury to have enough spectrum available for a hard frequency reuse – and even if they do have multiple carriers at hand, they may consider reusing all for both large and small cells for capacity maximization.

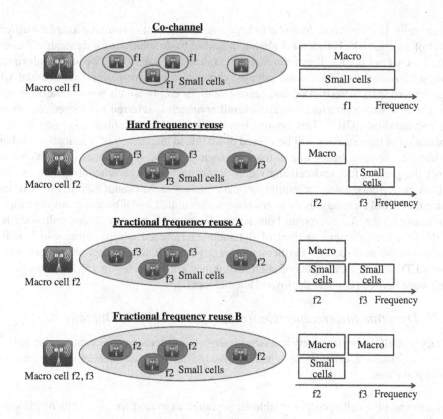

Figure 8.10   Frequency reuse options.

An alternative option is *fractional frequency reuse*, where some resources are reused by all layers, while others are dedicated to only one layer. In one variant of such a configuration, macro cells only use a frequency f2, but small cells may use frequencies f2 and f3. The decision on which frequency to use for a particular terminal connected to a small cell may then depend on the terminal's location: If this is located close to the border between macro and small cell, it should only use resources within f3 such as not to harm macro cell perform-ance, and if it is located in the center of the small cell it may be assigned to frequency f2. An alternative configuration of fractional frequency reuse is one where a frequency is reserved for macro cell usage only, which is often referred to as an *escape carrier*. This concept is particularly interesting to avoid aforementioned interference constellation C, as UEs which are close to a CSG femto cell to which they cannot connect can simply be served by the macro cell and on the escape carrier, so that no interference problems arise [2]. All stated frequency reuse options are illustrated in Figure 8.10.

8.4.2   Dynamic Interference Coordination in the Frequency Domain

Most operators will not have the luxury to be able to dedicate certain frequencies to macro-only or small-cell-only usage. Hence, one will mostly see scenarios of co-channel deployment, in conjunction with more dynamic means of adjusting resource reuse amongst adjacent cells. One option to do this is to have a flexible extent of resource usage for different chunks of spectrum. LTE Release 8 already included basic solutions for dynamically coordinating co-channel inter-cell interference by exchanging load and scheduling information between base stations. Then, base stations can take this information into account when assigning terminals to resources, and leave resources empty which would otherwise suffer from or create strong interference. The overall approach is referred to as inter-cell interference coordination (ICIC). This approach inherits two main problems: on one hand, the coordination of base stations will be slowed down due to the latency of information exchange over the X2 interface, hence one in the end has to trade the interference coordination gain against the gain of fast and cell-individual frequency selective scheduling. On the other hand, an interference-aware scheduling of data transmissions cannot help to alleviate interference on control channels, as these are always transmitted by all base stations regardless of the resource usage. To overcome both issues, Release 10 has introduced enhanced ICIC (eICIC) as a time-domain method of dynamic interference coordination, which will be explained in the next section. Furthermore, carrier-based ICIC schemes are under study for coming LTE releases examples of such techniques are discussed in [3] for managing the interference between densely deployed Home eNodeBs.

8.4.3 Dynamic Interference Coordination in the Time Domain

The target of eICIC is effectively to extend inter-cell interference coordination also to downlink control channels. eICIC is designed to solve the downlink co-channel interference problems in two cases:

1. Macro and pico cell scenario: enable larger range extension for pico cells by effectively addressing interference issue B in Figure 8.8.
2. Macro and femto cell with CSG: improve macro cell coverage when the macro subscriber is close to the co-channel femto cell; that is, addressing before mentioned interference issue C in Figure 8.8.

The basic principle is that some subframes are partially muted by using Almost Blank Subframes (ABS). An ABS is a subframe where only common reference signals (CRS) and other mandatory common channels are transmitted, while no data transmission occurs. Thus, during time instances where a cell uses ABS transmission, it generates much less interference to its surrounding cells. Figure 8.11 illustrates the operating principle between macro and pico cell. A macro cell allocates ABS which can be used by the pico cell to serve UEs that are further away from pico cell. Macro and pico cells can both reuse other subframes, for example, pico cells can use these to serve UEs that are closer to the pico cell. The benefit of ABS is that pico cells can serve more UEs, while the drawback is that part of the macro cell capacity is lost. Figure 8.12 shows the application of ABS between macro and CSG femto cells. In this case, macro cell users can still successfully receive signals close to a femto cell, if femto applies ABS [2].

Enhanced ICIC requires that macro cells and small cells are synchronized and coordinate the ABS muting patterns. The ABS configuration can be dynamically modified using an X2 interface. It can be assumed that an X2 interface exists typically between macro and pico cells, while it will not be available between macro and femto cell. The ABS configuration in

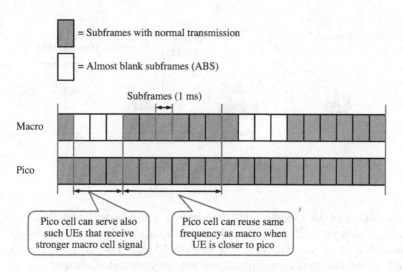

Figure 8.11 Enhanced ICIC between macro and pico cells.

a femto cell can be semi-static and it is controlled via the Operation and Maintenance (O&M) center, see Figure 8.13. New information elements have been added to the X2 application protocol (AP) in Release 10 to allow configuring of ABS muting patterns between eNodeBs. Although the new X2 signaling is standardized by 3GPP, the eNodeB algorithms are not standardized but are vendor specific. The typical adaptation of ABS muting patters between eNodeBs happens slowly compared to fast Radio Resource Management (RRM) within one eNodeB. The RRM actions take place with millisecond resolution, while ABS modifications would happen on the order of hundreds of milliseconds or several seconds. The ABS modifications cause only minimal control signaling load over X2.

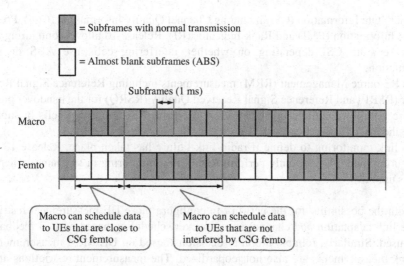

Figure 8.12 Enhanced ICIC between macro and Closed Subscriber Group (CSG) femto cells.

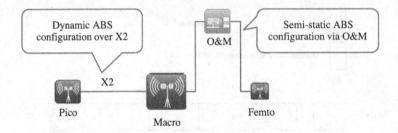

Figure 8.13 Almost Blank Subframe (ABS) configuration.

The efficient use of ABS requires that the terminal is also aware which frames are muted. ABS muting is defined in 3GPP Release 10 and in a backwards compatible manner so that also Release 8 UEs can co-exist with new Release 10 UEs. For this reason, ABSs are not completely muted, but some common channels are still transmitted, such as:

- Common Reference Signals (CRS),
- Common channels (synchronization channels, master information block, System Information Block (SIB) 1, paging, positioning reference signal) and also the PDCCH associated to SIB1/Paging, as well as Channel State Information Reference Signals (CSI-RS).

The periodical use of ABS naturally results in time-variant interference fluctuations. As an example, a UE served by a pico will experience significantly different channel quality depending on whether the macro is using normal transmission or ABS. Similarly, a macro UE close to a CSG Femto will also experience varying quality depending on whether the femto transmits ABS or normal subframes. It is therefore possible to configure Release 10 UEs with time-domain measurement restrictions:

- Channel State Information (CSI) including Channel Quality Information. (CQI), Precoding Matrix Information (PMI) and Rank Indicator (RI). Release 10 allows configuring UEs to report separate CSI depending on whether interfering cells use ABS or normal transmission.
- Radio Resource Management (RRM) measurements including Reference Signal Received Power (RSRP) and Reference Signal Received Quality (RSRQ) for the handover purposes. Release 10 allows configuring UEs to only perform RSRQ on pico-cells in subframes where the macro uses ABS.
- Radio link monitoring to define if radio link failure has taken place. Release 10 allows configuring macro-UEs to only perform Radio link monitoring in subframes where CSG femtos use ABS.

Thus, with the possibility for the network to configure those UE measurement restrictions, accurate link adaptation and channel aware packet scheduling based on CSI feedback can still be used. Similarly, reliable mobility decisions based on UE RRM measurements and radio link measurements are also not jeopardized. The measurement restrictions are configured with RRC signaling, and are valid only for intra-frequency measurements for connected mode. Configuration of UE measurement restrictions are not supported by Release 8 or 9 UEs. Therefore, legacy UEs may experience lower measurement performance in those networks where ABSs are utilized.

Another option is to use Multicast Broadcast Single Frequency Network (MBSFN) subframes. Only teh first control symbol and Common Reference Signals in the control region are transmitted in MBSFN subframes. MBSFN subframes cannot be configured in subframes 0, 4, 5, 9. Comparing ABS and MBSFN subframes: ABS can be used more flexibly, and generates interference only from Common Reference Signals. On the other hand, MBSFN allows sending uplink scheduling grants, but at the cost of interference from PDCCH. MBSFN is supported by Release 8 UEs.

8.4.4 Dynamic Interference Coordination in the Power Domain

In addition to dynamic interference coordination in frequency or time domain, one can also use modified power control rules to alleviate the impact of interference in heterogeneous deployments.

For example, interference problem A in Figure 8.8, where a small cell is subject to strong uplink interference from a terminal which is assigned to a macro cell, can be alleviated to a certain extent by reducing the terminal transmit power in an adaptive way. Similarly, the uplink interference caused by many femto UEs to nearby macro cells can be reduced. In both cases, the rule for UE transmission power can be stated as:

$$P = 10 \cdot \log_{10}(PRBs) + P_0 + \alpha \cdot PL + \Delta MCS + Closed\_loop \qquad (8.1)$$

Here, $PRBs$ is the number of scheduled Physical Resource Blocks (PRB), P_0 is normalized transmit power density, α is the pathloss compensation factor, PL is pathloss and ΔMCS is the quality requirement for the selected modulation and coding scheme. If α equals 1, then the pathloss is fully compensated by the power control. Fractional power control is used if α is below one.

For achieving the best uplink performance for co-channel macro and pico cases, it is recommended to use different P_0 settings for the macro- and pico-layer, respectively, as well as different settings of α. As a rule of thumb, the offset in P_0 between the two layers shall be on the same order of magnitude as the difference in downlink transmit power levels between macro and pico, and α shall be slightly smaller for the pico-layer. However, the optimal choice of uplink power control parameters depends on many factors in practice, and therefore needs to be adjusted accordingly. Here, options include using the closed loop component of the power control mechanism, as well as auto-tuning of power control parameters to maximize the desirable key performance indicators for the uplink. In general, one has to be aware that power control modifications always lead to trading the performance of one cell against the other, so that the interference coordination parameters will also depend on various considerations of the mobile network operator.

8.5 Performance Results

In this section, we now provide performance results for particular HetNet scenarios to which certain traffic steering and interference management solutions are applied.

8.5.1 Macro and Outdoor Pico Scenarios

The performance gain from using a co-channel deployment of macro cells and outdoor pico cells is summarized in this section. The performance results are obtained from system level simulations following the 3GPP evaluation methodology specified in [4]. The network topology is composed of a regular hexagonal grid of three-sector sites, complemented by one to ten outdoor deployed pico cells per macro cell area – that is total 3 to 30 pico cells per three-sector macro base station. The macro sites use 46 dBm transmit power and antennas of 14 dBi gain, resulting in an Equivalent Isotropic Radiated Power (EIRP) of 60 dBm. The pico cells use 30 dBm transmission power and omnidirectional antennas with 5 dBi gain, that is 35 dBm EIRP. A downlink 2×2 MIMO configuration is assumed with dynamic rank adaptation. A non-uniform UE distribution is assumed, with higher spatial density of UEs in the vicinity of the pico cells to model a scenario where the picos are deployed in hotspot areas corresponding to scenario 4b in [4].

The normalized downlink throughput gain from installing pico-eNodeBs is summarized in Figure 8.14. Normalization is performed relative to the case with no pico cells. Co-channel deployment without eICIC and without Range Extension (RE) is assumed. It is observed that simple co-channel deployment of macro and pico results in significant gains. As expected, the performance gain increases with number of picos, resulting in a significantly higher experienced end-user throughput. The user throughput increases more than three times when installing 10 pico cells per macro cell. Thus, the increase in user throughput is lower than the increase in number of available cells, since adding the cells also generates more interference.

By applying eICIC, the system performance can be further improved as this enables off-loading more users to the pico eNodeBs by using larger RE for those small cells. The larger RE is made possible by reducing the macro cell interference via periodical use of ABS. Figure 8.15 shows the percentage of users served by the pico eNodeBs versus the RE offset. The percentage of pico-users clearly depends on two factors; the number of picos per macro cell area as well as the RE offset. Increasing the RE offset significantly shift the balance so more users are offloaded from the macro-layer to the picos. As an example, applying 10 dB

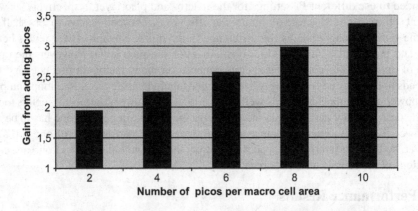

Figure 8.14 Normalized UE throughput performance gain from deploying pico eNodeBs as compared to macro-only scenario. No eICIC or range extension used.

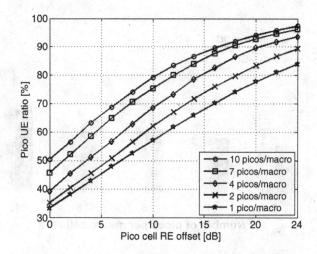

Figure 8.15   Percentage of pico-UEs versus the pico RE offset.

RE offset, the percentage of pico-UEs is increased from 40% (0 dB RE) to 70% for cases with 4 picos per macro cell area.

Figure 8.16 shows the 5 percentile throughput performance for a scenario with four picos per macro cell area. The performance is reported versus the assumed RE offsets for cases with different ratios of ABS at the macro-eNodeBs. It is observed from these results that the best performance is achieved by using ABS in four out of every eight subframes and 14 dB RE. If the RE offset is increased beyond 14 dB (for the considered case) the performance

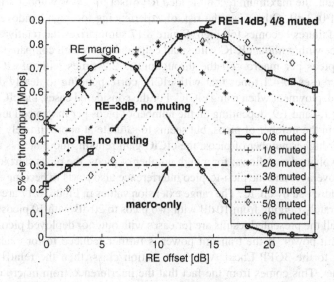

Figure 8.16   Performance versus the assumed pico-layer RE offset for different ABS muting ratios.

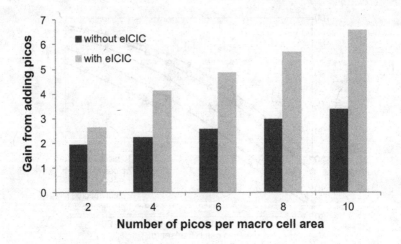

Figure 8.17 Relative eICIC performance for different number of picos per macro cell area (2 = double data rate).

decreases as this means offloading too many users to the picos. Similarly, the percentage of subframes configured as ABS in the macro-layer also influences on the overall network performance; if too many subframes are ABS, there are two few resources for the remaining macro-users, and if there are too few ABSs, the interference reduction for the picos becomes insufficient. Thus, for efficient operation of eICIC, it is important to have a coordinated setting of ABS ratio and RE offset, as supported by the Release 10 specifications via X2 signaling.

Notice that Figure 8.16 also includes a curve for the case with no ABS at macro. Here the performance is maximized for a RE of 6 dB. However, taking typical handoff margins of 3 dB into account, the maximum recommended RE offset for cases without ABS at macro is on the order of 0–3 dB. Otherwise, the risk of experiencing too high handover failure rates and radio link failures becomes too high. Figure 8.17 summarizes the relative user throughput performance with/without eICIC with small cells and macro cell compared to macro only case. With 4 picos per macro cell the throughput is 2.2 times without eICIC while the throughput increases to 4.1 times that with eICIC corresponding to nearly double benefit from pico cell deployment when using eICIC. For the considered cases, the eICIC gain varies between factor 1.4 and 1.9, depending on the number of picos per macro cell area. The eICIC gain increases in the number of picos, but seems to saturate at approximately four picos per macro cell. For higher number of picos, the eICIC gain remains more or less the same. The latter can be explained as follows; for a high number of picos, the picos start to have a fairly good overall coverage, and the pico-to-pico interference also starts to become more relevant, which is not addressed by eICIC. The range extension values in Figure 8.17 are optimized for each case separately ranging from 10 dB with two picos to 20 dB with 10 picos.

Notice that all the presented results are for cases with outdoor deployed pico-eNodeB with 30 dBm transmit power. If the transmit power is further reduced to for example, 24 dBm corresponding to the 3GPP Local Area base station class, then the relative eICIC gain increases further. This comes from the fact that the interference from macro cells becomes relatively stronger, and therefore there are larger benefits from using ABS as the macro-layer.

Similar, if the small cell transmission power is increased to typical micro cell levels, the eICIC gain is reduced. However, although the relative performance gain from eICIC is improved by reducing the transmission power of the small cells, the absolute experienced user performance is typically improved as the transmit power is increased.

For cases with indoor deployed picos, the eICIC gain is marginal: Mainly because indoor picos are installed to serve only the indoor users, which is typically possible without using large RE values. Applying large RE offsets for indoor picos and ABS at the macro to have the picos serve more outdoor users is often found to be a suboptimum solution.

The UE categories also have an influence on the performance benefits of using eICIC. As discussed earlier, Release 10 terminals support configuration of measurement restrictions for CSI, RRM, and RLM. Secondly, for achieving the maximum performance of eICIC, the terminals need to have advanced receiver capabilities with interference suppression of residual interference from ABS such as CRS interference. The presented eICIC results in this chapter were obtained under the assumption of such terminal support. Thus, for cases with legacy UEs without tailored eICIC support, the eICIC performance gain is likely to be lower.

8.5.2   Macro and Femto Scenarios

Co-channel deployment of macro and CSG femto cells is one of the most challenging cases from an interference point of view. Especially the downlink is challenging as the CSG femtos may cause macro-layer coverage holes, introduced as interference scenario C before. The aforementioned effect is now investigated by means of system level simulations, using the so-called dual-stripe scenario specified by 3GPP in [4]. This scenario consists of a regular three-sector macro-layer as well as multi-floor apartment building structures with indoor deployed CSG femtos, each having a transmit power of maximum 20 dBm. The building structure consists of two parallel building blocks, each having 20 apartments per floor – therefore the name 'dual-stripe scenario'. Inside each apartment with an active femto cell, it is assumed that there is a UE with matching CSG ID present. In addition to those UEs with configured CSG white lists, 80% of all macro UEs are randomly placed indoor and without any configured CSG ID, hence not being able to connect to any of the femto cells. A 10 MHz system bandwidth is assumed for the results presented in the following.

The first set of results in Figure 8.18 shows the macro-layer outage probability for a downlink target data rate of 200 kbps. A scenario with fairly high density of femto cells is simulated, assuming 48 active femtos per building. Results are shown for the case where all femtos transmit at 20 dBm or at 10 dBm, as well as cases with eICIC enabled and 20 dBm femto power. For the former eICIC case, it is assumed that the femtos use ABS in one out of eight subframes. Given the results in Figure 8.18 it is observed that without any interference management actions, co-channel deployment of macro cells and CSG femtos can indeed result in macro-layer outage. Reducing the femto cell power control brings additional benefits as the outage probability is further reduced. However, in order to avoid macro-layer coverage holes, some resource partitioning between macros and femtos is required. Here, eICIC is one of the options, which clearly brings additional benefits. Notice here that the eICIC results are obtained under the assumption of time-synchronization between macro and femtos and Release 10 UEs supporting configuration of RRM, RLM, and CSI measurement restrictions.

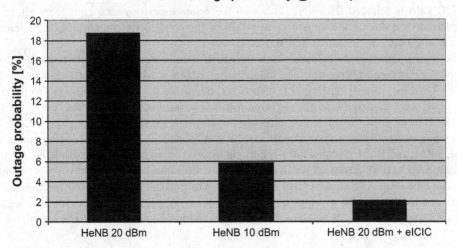

Figure 8.18 Macro-UE outage probability for 200 kbps depending on the Home eNodeB (femto) configuration.

An example of the cumulative distribution function (cdf) of the experienced end-user throughput is illustrated in Figure 8.19. Here, it is observed that the UEs being served by the femto cell are in general experiencing much higher throughput than the users on the macro-layer. Mainly because the femto-UEs are much closer to their serving cell and can be scheduled over the full bandwidth in every TTI. Macro-UEs, on the other hand, are further away from their serving cell and are typically multiplexed with many other macro-UEs, leaving less base station resources per user. However, with the introduction of femto cells, some of the users are offloaded to the femtos, resulting in a gain for the remaining macro-UEs

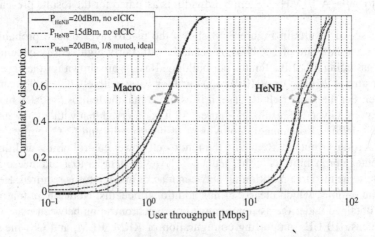

Figure 8.19 Macro and Home eNodeB (femto) UE throughput distribution [5].

as there are fewer users to share the available scheduling resources. Thus, the introduction of femto cells does provide end-user throughput benefits for all users. For the considered example, the femto-UEs are experiencing data rates of 10–30 Mbps, and hence the femto-UE throughput is comparable to the typical average aggregated macro-cell throughput. However, depending on the available backhaul connection for the femtos, this may actually become the limiting factor for the femto-UEs. Unlimited femto backhaul was assumed in these simulations.

8.6   Local IP Access (LIPA)

The LTE network architecture is designed for the centralized gateways where the operator typically only has one or a few gateways. That architecture makes sense for the Internet access because the number of Internet peering points is limited. Different architecture, however, may be needed for the small base stations to allow access to the local content. The local access would be practical for accessing corporate intranet information or accessing a home network over LTE radio. This section considers the Local IP Access (LIPA) with LTE radio.

LIPA function enables a UE to access directly enterprise or residential network without user plane data travelling first to the centralized gateway. LIPA functionality would make sense together with Home eNodeB (femto). A Local gateway (LGW) is co-located with a femto for the local access. The Local gateway has a subset of P-GW functionalities. The LIPA architecture is illustrated in Figure 8.20. A Home Subscriber Server (HSS) includes information for each APN and for each subscriber whether local access is allowed or not. If local access is allowed, the MME selects the LGW IP address enabling a direct user plane path between the LGW and the Home eNodeB. It is also possible to have multiple PDN connections: one to the macro network PGW and another one to the LGW. An S5 interface is used for controlling the LGW.

LIPA requires the following updates to the network elements.

- **HSS:** LIPA permission flag is given per subscriber per APN;
- **S1 interface:** LGW address is provided by Home eNodeB in initial NAS (Network Access Stratum) messages;
- **Home eNodeB:** Directing the LIPA bearers via the LGW and deleting LIPA bearers after handover;
- **MME:** Decide if LIPA can be provided based on HSS information;
- **LGW:** A few network functions with a subset of PGW functions.

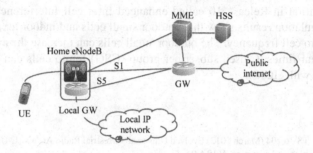

Figure 8.20   Local IP access (LIPA) architecture.

Note that LIPA does not require any changes to UE nor to the macro network gateways.

The operator's benefits from LIPA are mainly offloading the traffic from operator core network. The direct local access can also be considered as a new service for the customers. The local access can enhance the application area of LTE radio to the wider use in residential and in enterprise domains.

The first set of LIPA functions are defined in 3GPP Release 10. No mobility is supported in Release 10. If UE moves outside of Home eNodeB coverage are, LIPA connection is deactivated. A new connection will be re-established then to the macro gateway with a new IP address. Therefore, Release 10 solution is mainly designed for residential scenarios with one Home eNodeB per apartment. Release 11 included the mobility between multiple Home eNodeBs which would allow also more efficient use for larger enterprises.

8.7 Summary

The growth rate of data traffic drives the small cell rollouts because the macro cell capacity will be exhausted even if more spectrum and more advanced features are used in the macro cells. The small base stations are also getting more attractive from the hardware and from the optimization point of view: the base station size gets smaller and the advanced Self-Optimizing Network (SON) algorithms make the configuration and optimization simpler. In short, small cells are now driven by the fast traffic growth and by the new products with easy installation.

An intelligent traffic management between different cell layers brings the best end user performance and maximum capacity. The traffic management needs to consider signal levels, loading, service requirements and mobility.

WiFi offloading is one attractive approach for carrying part of the traffic to save mobile network capacity. The latest WiFi offloading solutions can make WiFi as an integral part of the mobile network with automatic network selection, automatic authentication and integrated to the mobile core network.

A central challenge inherent in the operation of heterogeneous networks is the interference amongst cells. Small cells may be rolled out on different frequency bands than macro cells, or on the same frequency band in the case of a so-called co-channel deployment. The co-channel case provides the highest total capacity but requires more advanced features and algorithms for the mitigation of interference between macro cells and small cells. For this, LTE enables frequency domain and power domain coordination in Release 8 and time domain coordination in Release 10 called enhanced Inter-cell Interference Coordination (eICIC). The simulation results show that outdoor small cells and indoor femto cells can co-exist on the macro cell frequency. The outdoor small cells can increase the average data rate by a factor of four times. Closed subscriber group (CSG) femto cells can co-exist on the macro frequency with eICIC.

References

1. 3GPP, Tech. Spec. TS 36.104 (March 2012) Evolved Universal Terrestrial Radio Access (E-UTRA); Base Station (BS) radio transmission and reception, V.10.6.0.

2. Szufarska, A., Safjan, K., Pedersen, K.I. and Frederiksen, F. (2011) Interference mitigation methods for LTE networks with macro and HeNB deployments. IEEE Proc. Vehicular Technology Conference (VTC-fall), September 2011.
3. Garcia, L., Pedersen, K.I. and Mogensen, P.E. (2009) Autonomous component carrier selection: Interference management in local area environments for LTE-advanced. IEEE Communications Magazine, September 2009.
4. 3GPP Tech. Spec. TR 36.814 (March 2010) Evolved Universal Terrestrial Radio Access (E-UTRA); Further advancements for E-UTRA physical layer aspects (Release 9), V.9.00.
5. Wang, Y. and Pedersen, K.I. (2011) Time and power domain interference management for LTE networks with macro-cells and HeNBs. IEEE Proc. Vehicular technology Conference (VTC-fall), September 2011.

9

Relays

Harri Holma, Bernhard Raaf and Simone Redana

9.1 Introduction

LTE and LTE-Advanced radio access can provide very high data rates, even beyond 1 Gbps, but only when the signal conditions are favourable. If the signal is weak, for example indoors, the data rates will be substantially lower than the theoretical peak data rates. The initial LTE rollout uses large macro cells and will not be able to provide full coverage for high data rates. There is a clear need to push the LTE coverage. The coverage can be pushed by installing more macro base stations or more small base stations. The additional base stations improve both coverage and capacity. The challenge with new base stations is typically to find the suitable base station site locations and to organize the transport connections. Another option for coverage improvement has been the use of RF repeaters which boost the signal level, but also boost interference. The interference control with repeaters has been a challenge in the existing networks. 3GPP Release 10 brings one more option for improving the network coverage: relays. The idea in the relay node is to use the same LTE air interface also for the backhaul connection, which makes the relay installation simple as no separate transport solution is required. Furthermore, the relays will be smaller equipment that makes it possible to deploy them on lamp posts, for example. The relay node is designed to be backwards compatible with Release 8, both from the UE and from the core network point of view. The relay node looks like a Release 8 eNodeB towards UEs. The access and the backhaul links are separated either in time or in frequency domain which will prevent any self-interference issues. An overview of the relay concept is shown in Figure 9.1. The access link refers to the connection between UE and relay node. The backhaul link refers to the connection between relay node and donor eNodeB.

This chapter first describes the relay node physical layer, architecture and protocols, and radio resource management. The coverage and capacity benefits are shown with simulations and example cost calculations are illustrated. Section 9.2 presents the general relay overview, Section 9.3 the relay physical layer, Section 9.4 the relay architecture and protocols, and

LTE-Advanced: 3GPP Solution for IMT-Advanced, First Edition. Edited by Harri Holma and Antti Toskala.
© 2012 John Wiley & Sons, Ltd. Published 2012 by John Wiley & Sons, Ltd.

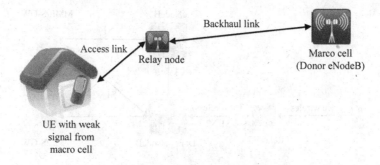

Figure 9.1 Overview of 3GPP relay concept.

Section 9.5 the radio resource management. The coverage and capacity gains in simulations are illustrated in Section 9.6. The future relay enhancements are discussed in Section 9.7 and the chapter is summarized in Section 9.8.

More information can be obtained from 3GPP documents. When relay standardization work started, 3GPP collected first requirements with relevant simulation assumptions in [1] and architecture alternatives in [2,3]. Subsequently the physical layer functionality was specified in [4], the enhancements to RRC and PDCP protocols in [5,6] and the enhancements to S1 and X2 interfaces in [7,8]. Performance requirements, the methodology to asses them and test procedures are collected in [9].

9.2 General Overview

The access link (Uu) between UE and relay node is based on Release 8 specifications. The backhaul link (Un) between relay node and the donor eNodeB has been standardized in Release 10 and reuses the protocols from the air interface (Uu). The radio protocols of the access link both for control plane and user plane are terminated at the relay node. The relay node looks like an eNodeB from UE perspective. Therefore, the relay node is backwards compatible and legacy Release 8 UEs can connect to Release 10 relay nodes. From the core network point of view, the relay node looks like an additional sector in the donor eNodeB. Similarly from a neighbouring eNodeB point of view, the UE looks as if it was connected directly to the donor eNodeB. The donor eNodeB acts as a proxy for the S1 and X2 interfaces relating to the Relay node, forwarding them towards core network and neighbour eNodeBs respectively. The architecture overview is shown in Figure 9.2 [3] and corresponds to the alternative two discussed in 3GPP [2].

The backhaul link has some enhancements compared to the access link [3]. The Radio Resource Control (RRC) layer [5] has the functionality to configure and reconfigure subframe configurations to adjust the resource split between backhaul and access links. The relay node can request a configuration from the donor eNodeB during the RRC connection establishment. The RRC layer has also the functionality to provide updated system information to the relay. There are some enhancements also on the user plane: the PDCP layer provides integrity protection for the user plane on the backhaul link because it is used to carry S1 and X2 control messages terminated at the relay node [6].

Physical layer enhancements on the backhaul link have become necessary in order to allow both access and backhaul link to share the same carrier [4]. This was implemented by

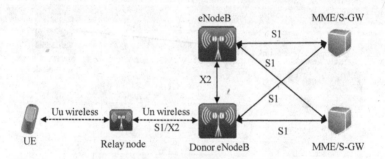

Figure 9.2   3GPP relay architecture and interfaces.

separating Uu and Un in time and caused relay specific extensions for both control and data signalling.

9.3   Physical Layer

When relay standardization work started, 3GPP collected first requirements with relevant simulation assumptions in [1]. Subsequently the physical layer functionality was specified in [4]. This functionality will be explained in the rest of this chapter.

9.3.1   Inband and Outband Relays

There are different options how to share the capacity between access link and backhaul link. The relay node needs to operate on four different unidirectional links: Both uplink and downlink with both UE and donor eNodeB. The inband relay uses the same frequency band for the access and the backhaul link while the outband relay uses different frequencies. The access and the backhaul links are separated in the time domain in case of inband relays. The difference between inband and outband relays is illustrated in Figure 9.3.

The frequency usage is shown in Figure 9.4. In case of inband relays, both backhaul and access link share the same frequency while outband relays use different frequencies for these links. Inband relays are more efficient from the spectrum usage point of view because both access and backhaul links are sharing the same spectrum. On the other hand, outband relays provide more capacity since the backhaul link does not eat from the access link capacity, at the expense of requiring more spectrum.

In case the same spectrum is available for both inband and outband relays they perform similarly in terms of capacity but inband introduces higher delay because of the time division

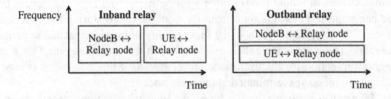

Figure 9.3   Principles of inband and outband relays.

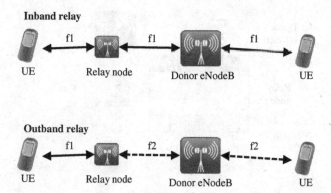

Figure 9.4 Frequency usage with inband and outband relays.

between backhaul and access links while outband requires enough isolation (frequency gap, filter, etc.) to cancel the adjacent frequencies interference between access and backhaul [10].

9.3.2 Sub-frames

The target in the inband relay design is to insulate the relay transmission and reception in the time domain to get around the interference problems of the traditional RF repeaters. The physical layer design must support the functionality that the relay transmission and reception happen at different times. It turned out that Release 8 already has in-built capability to support the time divided operation – via the so-called Multicast Broadcast Single Frequency Network (MBSFN) sub-frame. The Multimedia Broadcast Multicast Service (MBMS) system design was not completed in Release 8 but the physical layer support including the MBSFN sub-frame was already included into Release 8 in order to allow introduction of MBMS in a later release. MBSFN sub-frame includes short control section while most of the frame is targeted for multicast transmission. These MBSFN frames can nicely be used for the transmission from the donor eNodeB to the relay because UE does not expect any specific transmission from the relay during those sub-frames. The structure of the MBSFN sub-frame is illustrated in Figure 9.5 and the concept how to use it for relay transmission in Figure 9.6.

While the physical layer design for outband relays can be identical to the Uu design, for inband relays some changes are required compared to the access link. Physical Downlink Control Channel (PDCCH) of the access link uses Common Reference Signals (CRS) as reference and is transmitted over the whole bandwidth and is sent in the first few symbols of the sub-frame. This approach is required because many UEs signals are interleaved across

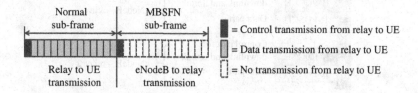

Figure 9.5 Use of MBSFN sub-frames.

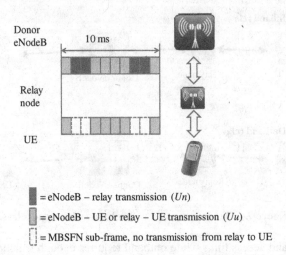

■ = eNodeB – relay transmission (*Un*)

□ = eNodeB – UE or relay – UE transmission (*Uu*)

⌐⌐ = MBSFN sub-frame, no transmission from relay to UE

Figure 9.6   Sub-frames for inband relays.

the entire bandwidth. The backhaul link includes a new control channel called Relay PDCCH (R-PDCCH) which is transmitted after the PDCCH, always starting at the third symbol. This new channel type is needed because the relay node may miss the first part of the sub-frame where PDCCH is transmitted because the relay node is still transmitting the PDCCH of the MBSFN sub-frame to its UEs. In order not to collide with PDSCH for users directly connected to the eNodeB, the R-PDCCH is localized on a few resource blocks in the frequency domain. R-PDCCH can also utilize dedicated Demodulation Reference Signals (DMRS), which can be sent specifically for each relay node by using beamforming. The data is carried by Relay PDSCH (R-PDSCH). R-PDCCH precoding may be different from data precoding on R-PDSCH. The new channels are illustrated in Figure 9.7 with an example case with two relays. 3GPP specifications support also more relays under each eNodeB.

R-PDCCH is used to dynamically assign resources to different relays within the semi-statically assigned sub-frames for the downlink backhaul data. R-PDCCH is also used to dynamically assign resources for the uplink backhaul data. The R-PDCCH may assign downlink resources in the current sub-frame and uplink resources in a corresponding sub-frame which happens typically 4 ms later. R-PDCCH is transmitted starting from an OFDM symbol within

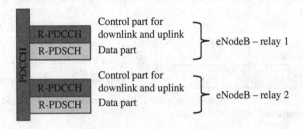

Figure 9.7   Control channel structure in backhaul link (Un).

the sub-frame that is late enough so that the relay has time to complete first the transmission of the control part of the MBSFN sub-frame towards its UEs, also taking the propagation delay and switching times into account.

9.3.3 Retransmissions

Hybrid Automated Repeat Request (HARQ) recovers from link errors by repeating incorrectly received packets. The retransmission delay in normal LTE is 8 ms. The uplink process is synchronous with fixed 8 ms retransmission delay while the downlink direction is asynchronous allowing also more than 8 ms delay. There can be in total eight parallel HARQ processes running. The same approach is not directly applicable to relays because MBSFN sub-frames can only be configured into six sub-frames out of 10 thus it is not possible to assign them strictly every 8 ms. The remaining four sub-frames carry essential information for the UE like synchronization signals, broadcast information regarding the cell configuration or paging messages. The solution with relays is to nonetheless keep 8 ms retransmission cycle and assign MBSFN sub-frames with a nominal repetition of every 8 ms: A sub-set of the set {0, 1, 2, . . . 7} is selected and an MBSFN sub-frame is configured, if the sub-frame-number module 8 lies within this set and if an MBSFN sub-frame can be configured at that particular sub-frame. Otherwise the retransmissions of all HARQ processes will be delayed till the next available sub-frame. If the selected subset contains only a single element the delay will be another 8 ms as illustrated in Figure 9.8, where the dashed line indicates the unavailable retransmission after 8 ms and the solid line the one taken instead after 16 ms. The more backhaul sub-frames are available and the more evenly they are distributed, the lower the additional delay is; for example only 2 ms as shown in Figure 9.9, which already achieves the lowest delay, despite the sub-set only containing four of the eight possible values. In the latter case the maximum of six backhaul sub-frames per 10 ms are configured. Depending on the number of backhaul sub-frames configured, the number of HARQ processes running in parallel will be less than the eight parallel processes on the Uu link: There is a single process in Figure 9.8, three processes in Figure 9.9, indicated by three pairs of arrows, and at most, six for the maximum backhaul configuration.

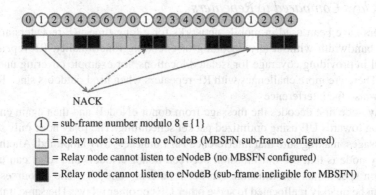

Figure 9.8 Retransmission in backhaul link for minimum backhaul configuration.

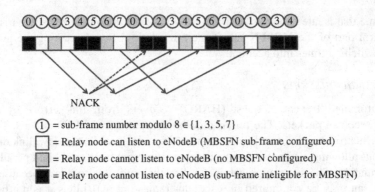

Figure 9.9 Retransmission in backhaul link when allocating half of the possible resources.

The end-to-end latency with relays is slightly higher than for direct UE – eNodeB connection.

- Two air interface transmissions, two scheduling and two processing delays are contributing to the total delay.
- Higher retransmission probability because of two radio links. This factor is partly compensated as the relay link can be assumed to be static and thus stable and typically there is ample capacity on the access allowing transmitting with some extra margin there.
- Higher retransmission delay sometimes in the backhaul link if MBSFN sub-frame is not available after 8 ms cycle.

The best case round-trip time with LTE is approximately 10 ms. The relay connection would roughly double the latency. In practice, however, the impact to the end user is not that dramatic since there are also other factors impacting the latency beyond radio, like transmission network and internet latency. Furthermore, relays allow higher throughput for cell edge users partially compensating the extra delay as a given data packet can be transferred quicker.

9.3.4 Relays Compared to Repeaters

RF repeaters have been used in mobile networks for a long time. RF repeaters amplify the whole RF bandwidth without any decoding or encoding functionality. RF repeaters have been useful in providing coverage for isolated locations, for example covering underground locations. There are more challenges with RF repeaters when used outdoors since RF repeaters amplify also the interference.

The relay node first decodes the message from donor eNodeB and then again encodes the transmission towards UE using optimized packet scheduling. The relay node only sends necessary messages, making sure that interference isn't unnecessarily amplified. Another benefit of the relay node is that the transmission between donor eNodeB and relay can use higher transmission speed than the transmission between relay and UE. The resources at donor eNodeB can be quickly reallocated to serve other UEs or other relays. The same transmission data rate must be used in both links in case of RF repeaters. The relay has also the benefit that

Table 9.1 Comparison between relays and RF repeaters

	3GPP Release 10 relay	RF repeater
Decoding and encoding	Yes	No, amplify RF including interference
Packet scheduling	Yes	No
Interference avoidance between transmission and reception	Yes, in time (inband) or frequency (outband) domain	Requires careful antenna planning
Possible to use different transmission speeds in the two links	Yes	No

there are no interference issues between its own transmission and reception because the different directions are separated in time or in frequency domain. The RF repeaters require careful planning and isolation of the antennas to avoid interference problems. The differences are summarized in Table 9.1 and in Figure 9.10.

In Figure 9.11 relay node and RF repeater spectral efficiencies are compared [11]. They are deployed at the donor eNodeB cell edge to improve user throughput (the deployment is shown later in Figure 9.22). The donor eNodeB transmit power is 46 dBm while the transmit power for relay/RF repeater is 30 dBm. The wireless backhaul link is assumed to be 16 dB better than the direct link assuming a UE in the same position of the relay/RF repeater. The SNR of the loop interference is assumed −5 dB that corresponds to a good isolation between the transmit and the receive antenna at the RF repeater. Figure 9.11 shows that in case at least five relay nodes are deployed they outperform repeaters for every quality of the link between relay/repeater and UE and for typical situations already for three relay nodes. We can further observe that if the access link SNR is good (user close to the relay), the gain from relay node is highest while if the access SNR is low (user far from the relay node) and there is only one relay node, the relay node can reduce the cell edge spectral efficiency compared to eNodeB only deployment. The reference cell edge efficiency without relays is (0.7 b/s/Hz); in this case the UE should be handed over directly to the eNodeB.

More relay nodes improve spectral efficiency because the links from the relays to their respective UEs can operate concurrently which is not possible for RF repeaters as shown in Figure 9.10.

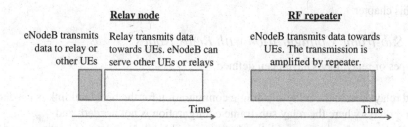

Figure 9.10 Inband relay compared to RF repeater.

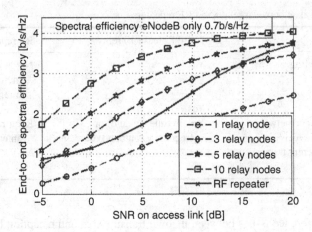

Figure 9.11 End-to-end spectral efficiency for relay nodes and RF repeaters, the cell edge spectral efficiency for eNodeB only deployment is 0.7 bit/s/Hz.

9.3.5 Relays in TD-LTE

Relays are also supported for Time Division Duplex (TDD) version of LTE (TD-LTE). For inband relays the same frequency can be used for uplink and for downlink, and for access link and for backhaul link. The configuration options are listed in Table 9.2. Let us look for example at relay configuration 2 which uses TDD configuration 1. That configuration allocates four sub-frames for downlink, four for uplink and two for switching. The switching sub-frames can also partly be used for data transmission; typically a large part of the switching sub-frame is used for downlink. The relay configuration 2 uses three sub-frames for the backhaul link and five for the access link. The switching sub-frames are also used for the access link. The uplink and downlink split in TDD must be the same in all cells. The split between backhaul and access links can be modified for different relays independently, as long as the interference between relay nodes is not causing problems, see Section 9.5.

9.4 Architecture and Protocols

When relay standardization work started, 3GPP collected the architecture alternatives in [2]. Subsequently the enhancements to RRC and PDCP protocols were specified in [5,6] and the enhancements to S1 and X2 interfaces in [7,8]. These enhancements will be explained in the rest of this chapter.

9.4.1 Sub-frame Configuration with Relay Nodes

Three types of relay nodes have been defined:

- inband relays where the relay sub-frame configuration for the backhaul link is needed;
- outband relays where the relay sub-frame configuration is not needed; and
- inband relays with adequate isolation between the backhaul and access antenna and hence the relay sub-frame configuration is not needed.

Table 9.2 Configuration options for TD-LTE relays

eNodeB – relay configuration	TDD Configuration	0	1	2	3	4	5	6	7	8	9
						Subframe numbers					
0		d	s	u	u	D	d	s	u	U	d
1		d	s	u	U	d	d	s	u	u	D
2	1	d	s	u	u	D	d	s	u	U	D
3		d	s	u	U	D	d	s	u	u	D
4		d	s	u	U	D	d	s	u	U	D
5		d	s	U	d	d	d	s	u	D	d
6		d	s	u	D	d	d	s	U	d	d
7	2	d	s	U	d	D	d	s	u	D	d
8		d	s	u	D	d	d	s	U	d	D
9		d	s	U	D	D	d	s	u	D	d
10		d	s	u	D	d	d	s	U	D	D
11	3	d	s	u	U	u	d	d	D	d	D
12		d	s	u	U	u	d	d	D	D	D
13		d	s	u	U	d	d	d	d	d	D
14		d	s	u	U	d	d	d	D	d	D
15	4	d	s	u	U	d	d	d	d	D	D
16		d	s	u	U	d	d	d	D	D	D
17		d	s	u	U	D	d	d	D	D	D
18	6	d	s	u	u	U	d	s	u	u	D

d = downlink in access link
u = uplink in access link
D = downlink in backhaul link
U = uplink in backhaul link

The relay node needs to indicate to the donor eNodeB that it is a relay (in order to differentiate from a UE) and the need for the relay sub-frame configuration in the RRC Connection Setup Complete message, as shown in Figure 9.12. The relay sub-frame configuration is provided by the donor eNodeB via dedicated signalling together with the system information update. The relay node cannot apply the normal system information acquisition procedure like a normal UE because the relay node acts as eNodeB for its UEs and hence it has to perform the system information acquisition procedure for these UEs. Only the SystemInformationBlockType1 (SIB1) and SystemInformationBlockType2 (SIB2) need to be provided to the relay node via dedicated RRC signalling, and the system information signalled to the relay node can be different from those that the donor eNodeB broadcasts to the UEs. The relay node does not apply system information validity timers on the stored system information received via dedicated signalling.

9.4.2 Bearer Usage with Relay Nodes

The operation and maintenance signalling for the relay node and the S1/X2 control signalling terminated at the relay node need to be carried over Un bearers. The Quality of Service (QoS) differentiation in Un interface uses the same QoS Class Identifiers (QCI) that are used also in S1 interface. The QCI values and their priorities are presented in [12]. The operation and maintenance signalling could use QCI 7 which is one of the non-guaranteed QoS classes.

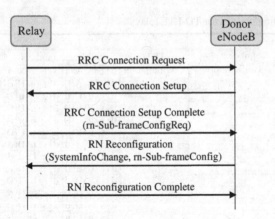

Figure 9.12 Relay node indication, Relay node sub-frame configuration request and Relay node reconfiguration.

The S1/X2 control signalling could use QCI 5 which is the highest priority bearer and used also for IMS signalling. The mapping between the bearers on the backhaul and access links is controlled by the relay node. An example of the bearer structure is shown in Figure 9.13.

9.4.3 Packet Header Structure in the Un Interface

The interface between the relay node and the donor eNodeB requires extra overhead on top of the user plane payload because of the GTP tunnel. For backhaul link, only the existing header compression will be used but no further enhancements are applied. This means that outer headers can be compressed but inner headers will not be compressed. There is some additional overhead caused by the IP header but it is considered very small, less than 10%, for typical web applications. The header compression would be beneficial if a large part of the traffic used small packets like VoIP connections. The packet structures in different interfaces are shown in Figure 9.14 for User Datagram Protocol (UDP) and Real Time Protocol (RTP) case like VoIP.

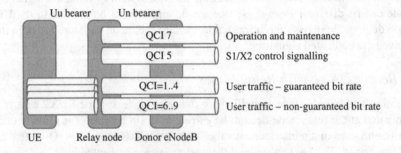

Figure 9.13 Bearers with relay node.

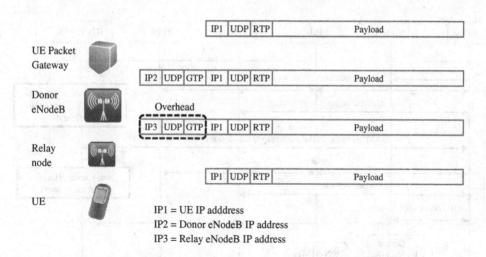

Figure 9.14   Packet header structure in relay interface for UDP/RTP case.

9.4.4   Attach Procedure

When a new relay node is installed at a site, it will automatically attach itself to the network. The attach procedure consists of two phases [3]. In the first phase the relay node creates RRC connection to the donor eNodeB and attaches itself as regular UE. During this phase the relay node obtains from Operation Administration Maintenance (OAM) the list of donor eNodeBs it is allowed to connect to and then the relay node detaches from the network. During the second phase the relay node connects to a donor eNodeB that is selected from the list acquired during the first phase. This ensures that the selected eNodeB does support relays, which would not be the case for legacy Release 8 and 9 eNodeBs. The attach procedure is shown in Figure 9.15.

During the second phase the relay node needs to inform the donor eNodeB that the attach is for a relay node because the donor eNodeB needs to select a relay node capable MME such that the MME can select the S-/P-GW which is collocated with the donor eNodeB. The MME retrieves from the Home Subscriber Server (HSS) the subscription data to verify that the relay node is allowed to connect; otherwise the RRC connection is released to prevent unauthorized relays to tap into the network.

9.4.5   Handovers

The handover between relay nodes can be done within one donor eNodeB or between two donor eNodeBs, see Figure 9.16. The handover procedure with relay nodes is similar to the handover between two sectors (intra donor eNodeB case) or the handover between two eNodeBs (inter donor eNodeB case). The downlink data arriving in the source relay node is forwarded from the source relay node to the target relay node during the handover process. The forwarding enables seamless handover without any packet losses. The forwarding is applied for inter donor eNodeB case but also for intra donor eNodeB case in Release 10 because of simplicity.

Release 10 supports only fixed relay nodes. The mobility of relay node itself is part of Release 11.

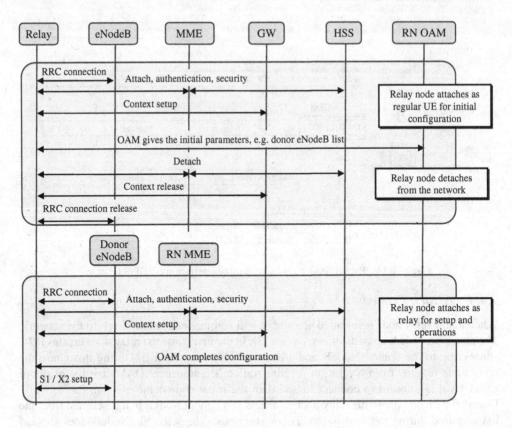

Figure 9.15   Relay node attach procedure.

9.4.6   *Autonomous Neighbour Relations*

The deployment of relay nodes may create new neighbour relationships and hence require to setup new X2 interfaces. An example case is illustrated in Figure 9.17. In case there are no relays, eNodeB1 has existing X2 interfaces to eNodeB2 and eNodeB3 but not with eNodeB4, since eNodeB1 and eNodeB4 do not have overlapping coverage areas. Instead there is a coverage hole in between. In order to cover this hole, a new relay node is deployed and connected to eNodeB1. This relay node has an overlapping cover- age with eNodeB4. Therefore, X2 interface is required between the relay node and eNodeB4. Since the relay node sets up the X2 interface with the donor eNodeB1 during the second phase of the attach procedure, this new X2 interface is created via donor eNodeB1 and hence requires the setup of an X2 interface between donor eNodeB1 and eNodeB4. Therefore the availability of the X2 interface between the relay node and eNodeB4 depends on the availability of the X2 interface between donor eNodeB1 and eNodeB4 while the use of the X2 interface for UE handover between relay node and eNode4 is configured by the relay node OAM. If the relay node is not allowed to use the X2 interface then the S1 interface has to be used.

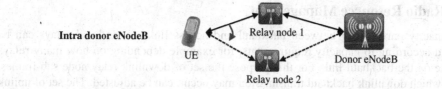

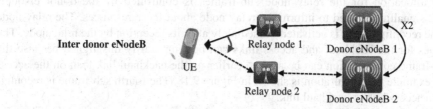

Figure 9.16 Intra and inter donor eNodeB handover cases.

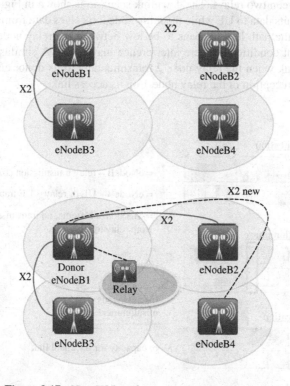

Figure 9.17 New X2 interface created via donor eNodeB1.

9.5   Radio Resource Management

The capacity partitioning between backhaul and access links for in-band relays can be adjusted according to capacity requirements, for example depending on how many relays compete on the backhaul link. For this purpose the set of downlink relay node sub-frames, during which downlink backhaul transmission may occur, can be adjusted. The set of uplink relay node sub-frames, during which uplink backhaul transmission may occur is implicitly derived from the downlink relay node sub-frames using the HARQ timing relationship. The resource allocation for the relay node sub-frames is controlled by the donor eNodeB and RRC signalling is used to inform the relay node about these resources. The relay node sub-frame reconfiguration is activated immediately after its reception by the relay node. The sub-frames for the backhaul and access links can be temporarily misaligned because the new sub-frame configuration can be applied earlier on the backhaul link than on the access link. An example reconfiguration is shown in Figure 9.18. The fourth sub-frame is reconfigured from access link to backhaul link.

The eNodeBs in LTE FDD operation do not need to be synchronized in the time domain. The time synchronization is only required for TDD operation, for MBMS transmission and may be needed for the location services. It can happen that one relay has access link transmission running at the same time when another relay receives data on the backhaul link if those two relays are connected to different eNodeBs. Even if neighbouring eNodeBs are synchronized, they may still use different sub-frame configurations making this interference scenario possible between two relays. The downlink scenario is shown in Figure 9.19 where the relay node 1 transmits data to UE while the relay node 2 receives data from its donor eNodeB simultaneously. If the path loss happens to be low between the relay nodes, for example in case of line of sight conditions, severe interference may result. A similar interference case can happen in uplink when the relay node 2 transmits data to its donor eNodeB and causes interference to the reception of the relay node 1 on its access link.

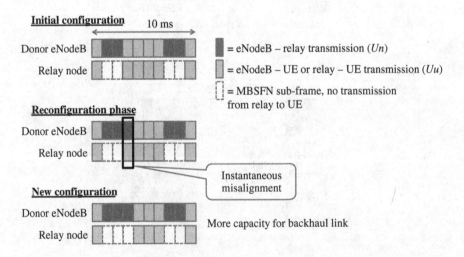

Figure 9.18   Reconfiguration of capacity between backhaul and access links.

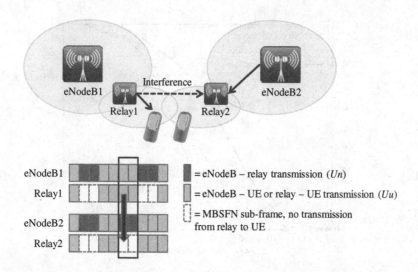

Figure 9.19   Interference scenario in downlink between two relay nodes.

The interference is more severe if 10 relay nodes are deployed per donor eNodeB compared to 4 relay nodes deployment scenario and it is more significant for higher inter-site distance (ISD) of 1732 m as compared to 500 m [13]. In Figure 9.20 different coordination strategies are compared for inter-site distance between donor eNodeBs of 1732 m and 10 relay nodes deployment. Full coordination refers to the case where neighbouring donor eNodeBs are synchronized and apply the same relay node sub-frame configuration. Intra-site coordination refers to the case where the same relay node configuration is applied in the cells of the same donor eNodeB while in case of no coordination different relay sub-frame configurations are even applied in cells belonging to the same donor eNodeB. We can observe that directional antennas screen out interference from relay nodes of neighbouring donor NodeBs efficiently and hence coordination is needed only within the site, without requiring synchronization amongst eNodeBs.

The relay nodes support only a lower maximum transmission power in downlink compared to donor eNodeBs and hence have a smaller coverage area and thus are typically only serving a few UEs. As a consequence, there is an inefficient use of radio resources because of the low load in the relay node cells and the still high load in the donor eNodeB cell. The cell borders between the donor eNodeB and the relay nodes can be adjusted biasing the cell selection and the handover thresholds, as shown in Figure 9.21. The UEs handed over from the donor eNodeB to the relays have more radio resources available that compensate the lower experienced SINR [14,15]. More detailed discussion about the co-channel deployment of macro cells and small cells can be found in Chapter 8.

9.6   Coverage and Capacity

Two deployment scenarios are considered where 4 relay nodes are deployed in one tier at the cell edge as well as 10 relay nodes which are deployed in two tiers at the cell edge, as shown in Figure 9.22. The assumed power levels are 46 dBm for the eNodeBs and 30 dBm for the relay nodes.

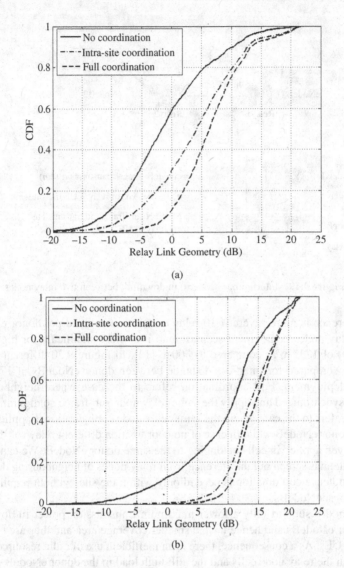

Figure 9.20 Relay link geometry for (a) non-directional and (b) directional backhaul antenna on the backhaul link and for different coordination levels of relay node sub-frame configurations.

9.6.1   Coverage Gain

The relay node deployment is considered in Release 10 mainly for coverage extension where relays are best deployed at (close to) the cell edge where users suffer from poor throughput and having a relay closer to these users can significantly boost their performance and thus the coverage.

In case the current eNodeB deployment does not meet the throughput requirements for the users located at the cell edge (Figure 9.23a), the traditional solution consists of increasing the density (Figure 9.23b) by deploying more eNodeBs. As alternative solution small nodes,

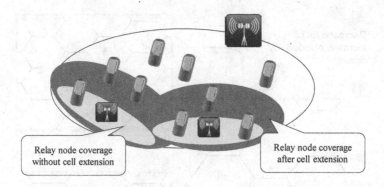

Figure 9.21   Relay node cell extension.

like pico nodes or relay nodes, can be deployed at the cell edge (Figure 9.23c) such that the performance of the cell edge users is the same as increasing the density of eNodeBs. We define the scenarios in Figure 9.23b) and Figure 9.23c) as isoperformance scenarios when the performance of the cell edge users is the same [16].

These two isoperformance scenarios can be characterized by the number of additional eNodeBs or additional relay nodes that need to be deployed. Basically the operator can decide whether to deploy one or the other and the relation between these numbers is called the exchange ratio between eNodeBs and relay nodes that provide the same performance in terms of cell edge throughput. Relay nodes are considered with in-band or with ideal backhaul, the latter is similar to pico nodes deployment or outband relays assuming sufficient spectrum is available for the backhaul link. Table 9.3 shows that relay nodes deployment with in-band and ideal backhaul have similar performance in scenario with inter-site distance (ISD) 500 m that is a similar number of relays can substitute an additional eNodeB. For inter-site distance of 1732 m the cell edge performance is the same when relays with in-band and ideal backhaul are deployed meaning relays are practically as efficient as picos and have the advantage not to require separate transport solutions [17].

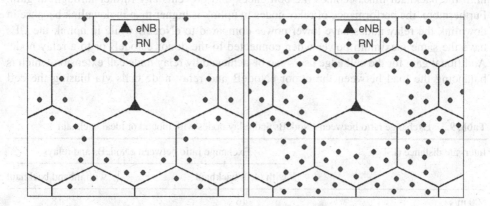

Figure 9.22   Relay node deployment in one (4 relays) and two (10 relays) tiers.

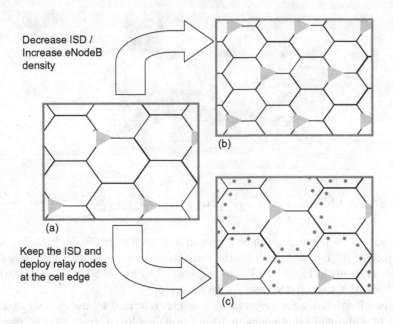

Decrease ISD /
Increase eNodeB
density

(b)

(a)

Keep the ISD and
deploy relay nodes
at the cell edge

(c)

Figure 9.23 Relay node deployment or increase of eNodeBs density to meet the performance requirements of cell edge users.

9.6.2 User Throughput Gains

We analyse the cell edge and average cell performance in relay deployment both in downlink and uplink. Results are relative to eNodeB only deployment both for inter-site distance of 500 m and 1732 m both in downlink [14,18–21] and uplink [15,22–27].

We can observe in Figure 9.24 that relay node deployment is particularly targeting scenarios with poor coverage; that is, 1732 m inter-site distance, where they show higher cell edge throughput gain. Relay nodes can also be used to increase capacity up to a certain level until the backhaul link becomes the bottleneck and prevents any further throughput gain. Furthermore, the performance of relay nodes in uplink is higher than in downlink because in downlink the relay nodes have lower power compared to eNodeB while in uplink the UEs have the same maximum power when connected to the donor eNodeB or to a relay node. Additional gain for the cell edge users can be achieved by relay node cell extension which is balancing the load between the donor eNodeB and relay node cells via biasing the cell

Table 9.3 Exchange ratio between eNodeBs and relay nodes with inband or ideal backhaul

Inter-site distance (ISD)	Exchange ratio between eNodeBs and relays	
	with ideal backhaul	with inband backhaul
500 m	15	18
1732 m	24	24

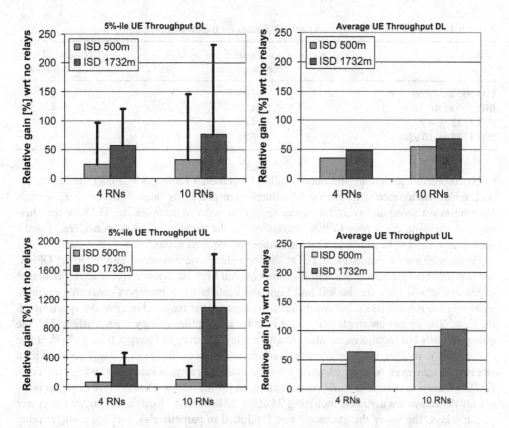

Figure 9.24 Cell edge and average cell throughput in downlink and uplink, the bars show the impact of the range extension on the cell edge throughput. Note different y-axis scale in fifth percentile uplink throughput bars.

selection and the handover thresholds and thus increasing the relay coverage. This is depicted as increase of the cell edge throughput shown by the bars in Figure 9.24.

These results have been obtained assuming the relay backhaul sub-frame configuration shown in Table 9.4. The number of relay sub-frames is selected according to optimization of the cell edge performance - (or equivalently the 5-percentile UE throughput).

9.6.3 Cost Analysis

The operator has two choices to increase the cell edge user experience, either adding additional eNodeBs or adding relay nodes as was explained in Section 9.6.1. Which one is more cost efficient depends on one hand on the required number of new nodes in both alternatives, that is, the exchange ratio. On the other hand it depends on the cost relation, more precisely on the total cost of ownership (TCO), which does not only include the cost of the nodes themselves but also for ancillaries like antenna mast, cabling, batteries, and so on and additional costs to plan the deployment, acquire the sites, civil work for opening up new sites, install the nodes and, in the case of eNodeBs, provide the transport connection. On top of these costs, which have to be paid once for installing the nodes (one-time costs), also recurring costs need

Table 9.4 Configuration of relay backhaul sub-frames for the different scenario

Scenario	Sector area compared to relay node coverage area	Number of relay sub-frames
ISD 500 m, 4 RNs	29%	2
ISD 500 m, 10 RNs	45%	4
ISD 1732 m, 4 RNs	44%	4
ISD 1732 m, 10 RNs	67%	6

to be considered to operate the nodes, called Operational Expenses (OPEX), including site rent, node maintenance (e.g. in case of failures) transport costs (leased lines or license costs for microwave links) and costs for power supply. In order to compare the TCO for the alternatives, one-time costs plus OPEX multiplied by the number of years considered for the deployment need to be multiplied by the number of required nodes.

Figure 9.25 shows examples of TCOs, distinguishing one-time costs and recurring OPEX, for two different scenarios, namely low site cost but high transport cost for example due to expensive leased lines on the left and high site costs but low transport costs, for example assuming microwave links and cheap licensing cost for the required microwave spectrum on the right. The former assumption is a best case for relay nodes as they are not affected by the transport costs but require more sites. A single relay site may be cheaper than a eNodeB site as the relay is smaller, but as multiple relays are needed in exchange of a single eNodeB total site rentals may even be more expensive for the multiple relays. Consequently high site costs and low transport costs is a worst case assumption for the relays. Furthermore different powers for the relays are assumed including 24 dBm, 33 dBm and 38 dBm, the higher the power of the relays, the lower the exchange rate (included in parenthesis), because a single relay can cover a larger area and thus less relays are required, but the higher the equipment cost per relay. The TCO of the eNodeB variant (leftmost bar) is normalized to 100% for both scenarios, consequently the columns for the different relays indicate how much cheaper (or sometimes more expensive) a relay deployment is. Further details can be found in [28].

Figure 9.25 in the top part, that is for best case assumptions, relays of any power are cheaper than eNodeBs, and more powerful relays have an even higher saving potential than low power relays as less relays are required. Recurring OPEX are comparable to one-time costs in this case.

For worst-case assumptions however, as shown in the bottom part, low power relays are even more expensive than eNodeBs, mainly due to the high site costs that need to be multiplied with the number of relays and which make the OPEX dominant over one-time costs. High power relays however, still achieve a significant saving.

These results assumed purely coverage limited scenarios where relays can be deployed effectively. Once the networks get capacity limited, relays can only be of limited help because eventually the backhaul link becomes the bottleneck.

9.7   Relay Enhancements

3GPP Release 11 is working on enhancing the relay functionality with moving relays. An example use case would be trains. High speed trains are currently deployed worldwide and providing the required services to the massive number of UEs that are moving with the train can be challenging. The carriages are well shielded with coated windows and therefore

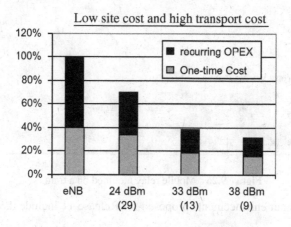

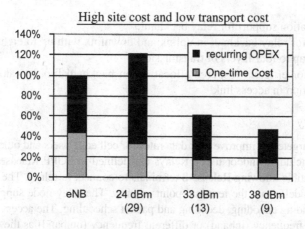

Figure 9.25 Cost comparison of five years TCO for eNodeBs and relays with different power assuming low site cost but high transport costs (top) and high site cost but low transport costs (bottom).

introduce a rather high penetration loss in the range of 20–30 dB. The high speed generates a Doppler frequency shift that causes frequency and channel estimation errors. Furthermore, the handover success rate of the connected UEs is lower compared to UEs moving with lower speed because of inaccurate or late neighbour cell measurements and excessive signalling; furthermore, frequent cell reselections drain UE batteries.

Mobile relays have gained attention recently as they are foreseen as a solution to combat issues that affect the UEs on board of the high speed trains [29], see Figure 9.26. A mobile relay is a base station/access point mounted in a moving vehicle that provides wireless connectivity service to end users inside the vehicle via an indoor antenna. The wireless backhauling connection to on-land network is provided via an outdoor antenna. The work item for mobile relays is beyond Release 11.

One potential enhancement is the capability to allow different air interface technologies on the access link: the wireless backhaul link uses the LTE air interface while GSM and HSPA may also be provided on the access link.

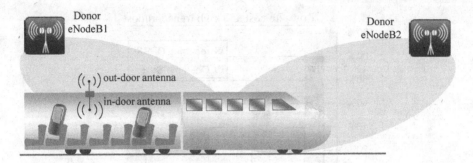

Figure 9.26 Mobile relay mounted on a train.

Other relay backhaul enhancements proposed for Release 11 include the following topics [30]:

- Carrier Aggregation support for relay backhaul.
- Improvements for relay backhaul in uplink and downlink with up to eight transmit antennas supporting up to four layers of transmission.
- Enhancement for efficient resource allocation on backhaul link by taking into account resource allocation in access link.

9.8 Summary

Relay nodes are targeted to improve user data rates for cell edge users and other users with poor radio coverage, including indoor users. Relays are defined in 3GPP Release 10 and they are backwards compatible allowing Release 8 terminals to connect to relays. The relay node looks like a normal eNodeB from the terminal point of view. The relay node supports full eNodeB functionality including encoding, decoding and packet scheduling. The access link in the relays can use the same frequency (inband) or different frequency (outband) as the backhaul link. In case of inband relays, the access and backhaul links are separated in the time domain to avoid any interference issues between the two links. The relay nodes are therefore clearly easier to deploy than traditional RF repeaters from the interference point of view. The resource allocation between access and backhaul links can be adjusted according to capacity requirements.

The simulations show that relay nodes bring a substantial improvement in the cell edge user data rates. The benefit is highest when the relay node can be placed close to the user thus providing good signal -to-noise ratio. The system capacity will also benefit if the quality of the backhaul link is good because it allows minimizing the capacity allocation for the backhaul and giving more capacity for the access link and the users served by the donor eNodeB. The relay node brings the largest benefit for the operator if the backhaul cost of the traditional eNodeB is high, site cost is low and the network is coverage limited.

References

1. 3GPP TR 36.814 (March 2010) Evolved Universal Terrestrial Radio Access (E-UTRA); Further Advancements for E-UTRA Physical Layer Aspects (Release 9).
2. 3GPP TR 36.806 (March 2010) Evolved Universal Terrestrial Radio Access (E-UTRA); Relay architectures for E-UTRA (LTE-Advanced).

3. 3GPP TS 36.300 (March 2012) Evolved Universal Terrestrial Radio Access (E-UTRA) and Evolved Universal Terrestrial Radio Access Network (E-UTRAN); Overall description; Stage 2 (Release 10).

4. 3GPP TS 36.216 (September 2011) Evolved Universal Terrestrial Radio Access (E-UTRA); Physical layer for relaying operation.

5. 3GPP TS 36.331 (March 2012) Evolved Universal Terrestrial Radio Access (E-UTRA); Radio Resource Control (RRC); Protocol specification (Release 10).

6. 3GPP TS 36.323 (March 2011) Evolved Universal Terrestrial Radio Access (E-UTRA); Packet Data Convergence Protocol (PDCP) specification (Release 10).

7. 3GPP TS 36.413 (March 2012) Evolved Universal Terrestrial Radio Access (E-UTRA); S1 Application Protocol (S1AP) (Release 10).

8. 3GPP TS 36.423 (March 2012) Evolved Universal Terrestrial Radio Access (E-UTRA); X2 application protocol (X2AP) (Release 10).

9. 3GPP TR 36.826 (March 2012) Evolved Universal Terrestrial Radio Access (E-UTRA); Relay radio transmission and reception (Release 9).

10. Góra, J. and Redana, S. (2011) In-Band and out-band relaying configurations for dual-carrier LTE-advanced system. Proceedings IEEE International Symposium on Personal, Indoor and Mobile Radio Communications (PIMRC).

11. Bou Saleh, A., Redana, S., Raaf, B. et al. (2009) Performance of amplify-and-forward and decode-and-forward relays in LTE-advanced. IEEE Vehicular Technology Conference (VTC) Fall.

12. Holma, H. and Toskala, A. (2011) LTE for UMTS, 2nd edn, John Wiley & Sons, Ltd, Chichester.

13. Bou Saleh, A., Bulakci, Ö., Redana, S. et al. (2011) A divide-and-conquer approach to mitigate relay-to-relay interference, Proceedings IEEE International Symposium on Personal, Indoor and Mobile Radio Communications (PIMRC), September.

14. Bou Saleh, A., Bulakci, Ö., Redana, S. et al. (2010) Enhancing LTE-advanced relay deployments via Biasing in cell selection and handover decision. Proceedings IEEE International Symposium on Personal, Indoor and Mobile Radio Communications (PIMRC).

15. Bulakci, Ö., Bou Saleh, A., Redana, S. et al. (2010) Enhancing LTE-Advanced Relay Deployments via Relay Cell Extension. International OFDM-Workshop (InOWo).

16. Beniero, T., Redana, S., Raaf, B. and Hämäläinen, J. (2009) Effect of relaying on coverage in 3GPP LTE. IEEE Vehicular Technology Conference (VTC) Spring.

17. Bou Saleh, A., Redana, S., Raaf, B. and Hämäläinen, J. (2009) Comparison of relay and Pico eNB deployments in LTE-advanced. IEEE Vehicular Technology Conference (VTC) Fall.

18. Bou Saleh, A., Redana, S., Raaf, B. and Hämäläinen, J. (2010) On the coverage extension and capacity enhancement of inband relay deployments in LTE-advanced networks. Journal of Electrical and Computer Engineering, doi: 10.1155/2010/894846.

19. Bou Saleh, A., Bulakci, Ö., Ren, Z. et al. (2011) Resource Sharing in Relay-enhanced 4G Networks - Downlink Performance Evaluation, European Wireless Conference (EW).

20. Bou Saleh, A., Bulakci, Ö., Redana, S. et al. (2011) Addressing Radio Resource Management Challenges in LTE-Advanced Relay Networks - Downlink Study, VDE 16. ITG Workshop on Mobile Communications.

21. Ren, Z., Bou Saleh, A., Bulakci, Ö. et al. (2012) Joint Interference Coordination and Relay Cell Expansion in LTE-Advanced Networks, IEEE Wireless Communications & Networking Conference (WCNC).

22. Bulakci, Ö., Redana, S., Raaf, B. and Hämäläinen, J. (2010) System optimization in relay enhanced LTE-advanced networks via uplink power control. IEEE Vehicular Technology Conference (VTC) Spring.

23. Bulakci, Ö., Redana, S., Raaf, B. and Hämäläinen, J. (2011) Impact of power control optimization on the system performance of relay based LTE-advanced heterogeneous networks. Journal of Communications and Networks, 13 (4), 345–359.

24. Bulakci, Ö., Bou Saleh, A., Ren, Z. et al. (2011) Two-step Resource Sharing and Uplink Power Control Optimization in LTE-Advanced Relay Networks, International Workshop on. Multi-Carrier Systems & Solutions (MC-SS).

25. Bulakci, Ö., Bou Saleh, A., Redana, S. et al. (2011) Uplink Radio Resource Management Challenges in LTE-Advanced Relay Networks, VDE 16. ITG Workshop on Mobile Communications.

26. Bulakci, Ö., Bou Saleh, A., Redana, S. et al. (2011) Flexible Backhaul Resource Sharing and Uplink Power Control Optimization in LTE-Advanced Relay Networks, IEEE Vehicular Technology Conference (VTC) Fall.

27. Bulakci, Ö., Awada, A., Bou Saleh, A. *et al.* (2011) Joint optimization of uplink power control parameters in LTE-Advanced relay networks, 7th International Wireless Communications and Mobile Computing Conference (IWCMC).
28. Lang, E., Redana, S. and Raaf, B. (2009) Business impact of relay deployment for coverage extension in 3GPP LTE-advanced, international workshop on LTE evolution. IEEE International Communications Conference (ICC).
29. RP-111377 (2011) Mobile Relay for E-UTRA, 3GPP TSG RAN#53, Fukuoka, Japan, September 13–16.
30. RP-111075 (2011) Improvements to LTE Relay Backhaul, 3GPP TSG RAN#53, Fukuoka, Japan, September 13–16.

10

Self-Organizing Networks (SON)

Cinzia Sartori and Harri Holma

10.1 Introduction

The target of Self-Organizing Network (SON) features is to improve the network quality and capacity and to simplify the network operations. SON algorithms are designed to provide better end user performance, lower cost of operations and higher network utilization. The need for SON emerges especially together with LTE since LTE introduces an additional radio network on top of existing GSM and HSPA. LTE and heterogeneous networks also bring more base stations. At the same time the network operating costs should be minimized. SON is one of the solutions to achieve these targets. SON features have been introduced in 3GPP together with LTE radio starting in Release 8 and the capabilities have been enhanced in further 3GPP releases. This chapter summarizes the SON features and focuses on the latest additions in the specifications.

10.2 SON Roadmap in 3GPP Releases

Multiple SON use cases have been specified by 3GPP for self-configuration, self-optimization and self-healing areas. In general self-configuration use cases are important at earlier phases when providing coverage whereas self-optimization and healing consider quality and capacity and they are more relevant when network usage increases.

The SON use cases are specified by three working groups according to their responsibilities: RAN2, RAN3 and SA5. RAN2 addressed only one use case: Minimization of Drive Tests (MDT) while other cases are defined by RAN3 and SA5.

While self-configuration features are mainly specified in 3GPP Release 8 and include Automated Configuration of Physical Cell Identity (ID) and Automatic Neighbour Relation

---

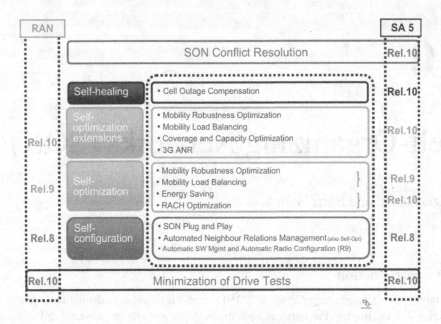

Figure 10.1 SON roadmap in 3GPP.

Function, self-optimization is the core content of 3GPP Release 9, where focus is on Mobility Load Balancing (MLB), Mobility Robustness Optimization (MRO) and RACH optimization. SA5 deals with self-healing features: Cell Outage Compensation (COC) is part of Release 10. Release 10 work started in 2010 and provides MRO and MLB enhancements and Minimization of Drive Tests (MDT).

Release 11 SON still focuses mainly on failure scenarios; that is MRO in inter-RAT and HetNet plus minimization of ping-pong events; avoiding ping-pongs will reduce the number of unnecessary mobility events (i.e. signalling) and decrease UE power consumption. The main focus of Release 11 MDT is QoS verification.

Energy Saving (ES) has been specified since Release 9 for Intra-LTE. However, an Energy Saving Release 10 Study Item extends ES to more scenarios, like the multi-RAT one; in addition mechanisms for waking up cells which are in 'energy saving' state have also been discussed. The Energy Saving Work Item is in Release 11.

SON overview in 3GPP is illustrated in Figure 10.1. More details about use cases can be found in Chapter 3 of [1].

LTE-Advanced features are introduced in Release 10: those include carrier aggregation, uplink and downlink MIMO enhancements, relay nodes and optimized interworking between cell layers in heterogeneous network. Self-configuration and Automatic Neighbour Relations are specified for relay nodes. While the SON use cases specified for macro apply nicely to small cells as well, SON work Item in Release 11 specifically addresses multi-layer scenarios to tailor SON use cases to heterogeneous networks even further.

10.3 Self-Optimization

10.3.1 Mobility Robustness Optimization

The general task of Mobility Robustness Optimization (MRO) is to guarantee good mobility, that is, proper handovers in connected mode. This is achieved by:

- Minimizing Radio Link Failures (RLF) and call drops; in many cases a connection can be re-established after an RLF before the call drops; or, in worst case, a mobility problem may lead to a call drop. Connection re-establishment is only possible inside LTE, and not possible if another target RAT is involved. Recovery via the core network can help maintaining user service if it can bear temporary degradation of quality.
- Minimizing unnecessary handovers and ping-pongs. Ping-pongs refer to repeated handovers between two cells within a short time. Unnecessary handovers and ping-pongs do not lead to RLF but many of those would lead to inefficient use of network resources and reduction of user throughput so that user perception may be affected as well.

MRO in Release 9 is specified for LTE only (both intra and inter-frequency), while Release 10 addresses some Inter-RAT challenges, such as unnecessary handovers. The full Inter-RAT MRO is a core topic of Release 11. An extensive description, scenarios and related simulations are included in [1].

The MRO idea is to automatically detect and correct errors in mobility configuration that lead to connection failures. This is enabled by failure reporting specified in Release 9 and later extended in Release 10; before Release 9 it was impossible to identify, in most cases, the cause of the failure, for example a coverage hole, too-late handover, too-early handover. MRO assumed most failures are due to mobility and provided a framework necessary to recognize the most relevant failure causes.

An important issue that was agreed when defining Release 9 MRO is that too late handover case (definition is given in Section 10.3.1.1) should be differentiated from a coverage hole, otherwise MRO could take wrong corrective actions and get the situation worse. This is in line with the fact that MRO tries to optimize the network, but of course it cannot do much in case of a coverage hole, no matter how the handover thresholds are changed. Coverage optimization is in the scope of MDT (Section 10.3.3).

MRO can be split into two subtasks, the root cause identification/evaluation and the actual correction of the mobility parameters. While 3GPP primarily deals with the root cause identification, parameters' correction is typically vendor specific.

10.3.1.1 MRO Release 9

Before introducing MRO enhancements in Release 10, a short overview of Release 9 is given. Mobility Robustness Optimization is described in [2] and addresses three types of handover problems:

- Too-late handover,
- Too-early handover,
- Handover to wrong cell.

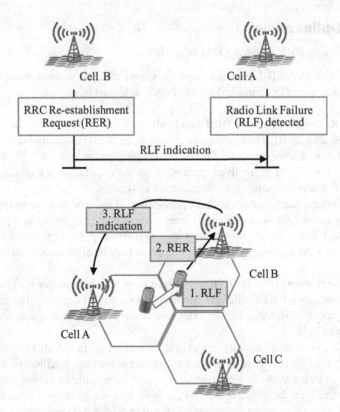

Figure 10.2   Detection of a too-late handover.

Too-late handover (Figure 10.2): A connection failure occurs in Cell A and the connection is re-established in Cell B.

A too-late handover happens when a connection failure occurs in a cell (Cell A) before the handover was initiated. The UE attempts to re-establish the radio link connection in the target cell (if handover was initiated) or in a cell which is not the source cell (if handover was not initiated) (Cell B). Figure 10.2 illustrates the mechanism: the UE suffers an RLF (1) when moving from Cell A to Cell B and sends a re-establishment request ('RER', 2) to Cell B. Cell B may request the RLF report with containing measurements before the RLF (not shown in the figure). Cell B informs Cell A about the detected RLF via RLF Indication (3). Cell A realizes that there was no recent handover of this terminal, so Cell A knows that it has created the 'too-late handover' problem.

Too-early handover: (Figure 10.3) A successful handover is executed from cell B to A, then very soon connection fails in cell A and is re-established in B.

The UE moves in Cell B through a coverage island of Cell A. It is successfully handed over to Cell A (1), but suffers an RLF shortly after (2) and sends the re-establishment request to Cell B (3). Cell B informs Cell A about the RLF via RLF indication (4). Cell A realizes that there was a recent handover shortly beforehand such that it had no chance to save the UE. So it will inform Cell B via handover report (5), that it has created a too early handover. The connection may fail also during the handover from B to A. However, since the procedure

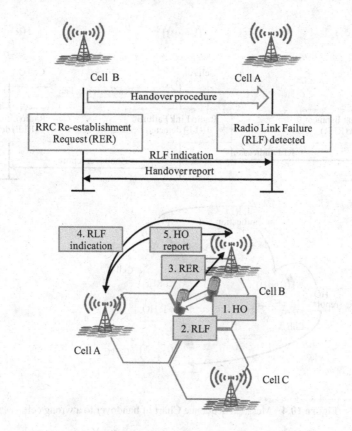

Figure 10.3 Message sequence chart of a too-early handover.

is not completed, the Cell B can recognize the UE when it requests re-establishment and no inter-eNodeB signalling is needed.

Handover to wrong cell (Figure 10.4): A successful handover is executed from cell C to B, then very soon connection fails in cell B and is re-established in A or the handover to B fails in the process): handover to wrong cell in B (i.e. cell C should have performed the first handover to cell B, not to A).

The UE moves from Cell C to Cell B, closely passing Cell A. It is successfully handed over to Cell A (1), but suffers an RLF shortly after (2), or if the procedure is not completed then handover fails. The UE sends the re-establishment request to Cell B (3). If the prior handover to A was completed, Cell B informs Cell A about the RLF via RLF indication (4). Cell A realizes that there was a recent handover shortly beforehand such that it had no chance to save the UE. So it will inform Cell C via a handover report, that it has created a handover to wrong cell. If the prior handover to A was not completed, the RLF INDICATION is sent directly to C, which can further recognize the problem (it still has the UE context).

Ping-pongs have not been addressed in Release 9 since they can be read from the UE history which is exchanged along with the UE context during the handover (Release 8 rudimentary solution). The UE history contains the time the UE has stayed in each cell. For a

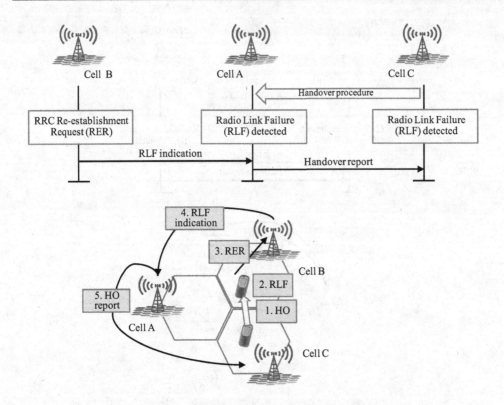

Figure 10.4   Message Sequence Chart of handover to a wrong cell.

ping-pong: 'Cell A – Cell B – Cell A', the UE history will tell Cell A that the UE was already in Cell A shortly before. Note that Cell A will not recognize the UE, since the previous C-Radio Network Temporary Identifier (RNTI) is lost.

10.3.1.2   MRO Release 10

The MRO solution defined in Release 9 is limited by the fact that the UE successfully re-establishes the RRC connection and is able to send measurements in the RLF Report to the eNodeB. If the re-establishment procedure is not successful and the UE goes to idle, the additional measurements contained in the RLF report are lost.

In general RLF report is lost in case the UE appears in an unprepared eNodeB. This may happen in a number of cases listed in [3]. According to the RRC specification, if the re-establishment fails or is not possible, the UE changes its state to RRC_IDLE and starts cell selection procedure. Once a suitable cell is found, the UE attempts RRC connection setup. The setup procedure is different from re-establishment, and the information provided from the UE is different, for example identification of the last cell that served the UE or the identification of the UE in that cell are not provided in RRC fresh setup procedure. Furthermore, there is no time guarantee for the establishment: due to NAS recovery, it is likely to happen soon after the connection failure, but a longer delay cannot be ruled out. In the latter case the

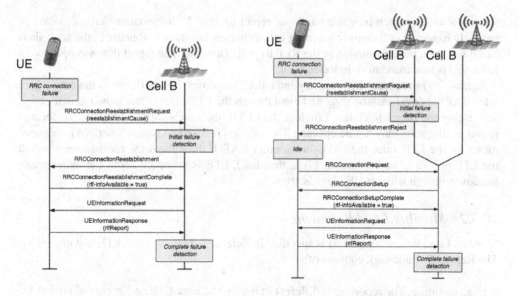

Figure 10.5 RLF Report retrieval after successful RRC-Reestablishment (left) and after RRC-IDLE transition (right).

reconnection may happen in significant distance from the cell that last served the UE. A typical scenario is the one of a large coverage hole.

In case of reconnection after idle mode, 3GPP decided for the UE-based solution; in Release 10 the RLF Report can also be sent after the UE went to idle mode [4,5]. The RLF report, if not sent, is stored in the UE for 48 hours. It shall also survive to state transitions and RAT changes (see Figure 10.5).

The use of RLF report after RRC_IDLE is twofold; it can be used to integrate MDT reports in the coverage optimization (MDT is addressed in Section 10.3.3) as well as in MRO use case. RLF Reports includes the following UE information:

- Cell measurement results from serving cell as well as available neighbours' cells at the time of the failure. In particular the Cell Identification (ID) for following cells, either Global Cell ID (E-CGI) or Physical Cell ID (PCI):
  - Cell ID where the failure happened;
  - Cell ID of the cell where the last handover command was received;
  - Cell ID of the cell to which the UE attempted to establish the connection after the failure.
- Time between last successful handover initialization and the failure.
- Available accurate location information from standalone Global Navigation Satellite System (GNSS) function at the time of the failure.
- Indicator on connection failure type, either radio link failure or handover failure.

A Release 10 UE shall send an indication about the existence of the RLF report also in case of a RRC Connection Reconfiguration for example after a handover. Once the network is informed about the failure, the eNodeB that received the RLF Report forwards it to the last serving eNodeB, which may use handover report to pass the information further in case of too early handover or handover to wrong cell. However, contrary to Release 9, the analysis is based solely on the information in the RLF Report. Therefore, the report that was optional in Release 9 is now mandatory in Release 10.

Release 10 introduces 'Unnecessary inter-RAT Handovers' as well; this is the case where inter-RAT handover may be triggered even though the LTE layer offers good enough coverage. Being the target RAT less efficient than LTE, the user's satisfaction may suffer and resource allocation may not be optimal. The solution is based on setting inter-RAT measurements for the LTE, once the UE is in the other RAT. If throughout the measurement period the LTE proved to be better than given threshold, LTE is informed about the unnecessary handover, though without UE identification.

10.3.2 Mobility Load Balancing

Mobility Load Balancing (MLB) is specified in Release 9, both for intra-LTE and inter-RAT. The Release 9 framework consists of:

- Load reporting. The reporting is different in the case of intra-LTE and in case of inter-RAT.
- Load balancing action. The load balancing in Release 8 and Release 9 concerns active user and therefore is based on handovers.
- Amending handovers so that load remains balanced.

Load reporting, or load information exchange, enables eNodeBs to exchange information about load level in cells and about available capacity. In case of load information exchange between LTE cells, the reporting is initialized by one eNodeB toward a peer eNodeB with the Resource Status Reporting Initiation X2 procedure. For inter-RAT information exchange RIM is used and the 'SON container' is defined to transport SON information over RIM (container defined in [6]). Release 10 only introduces minor enhancements to MLB:

- LTE: partial failure for LTE resource status reporting.
- Multi-RAT: multiple-cells and event-based reporting.

An example MLB case is illustrated in Figure 10.6. eNodeB A has an overload while the adjacent eNodeB B has lower loading. eNodeB A requests load information from eNodeB, and from other adjacent eNodeBs. eNodeB A finds the suitable candidate for offloading the traffic. eNodeB A proposes then modifications to handover parameters. The new parameters are activated when eNodeB acknowledges the changes.

10.3.3 Minimization of Drive Tests

Operators normally perform drive tests to monitor their mobile network and collecting network statistics. The current testing methodology comprises a number of steps and is controlled manually. The process starts with drive tests, which are usually performed by skilled field-engineers. Those specialists measure the network in given areas, for example urban or suburban sites, typically using a special car or van which is adequately equipped with test

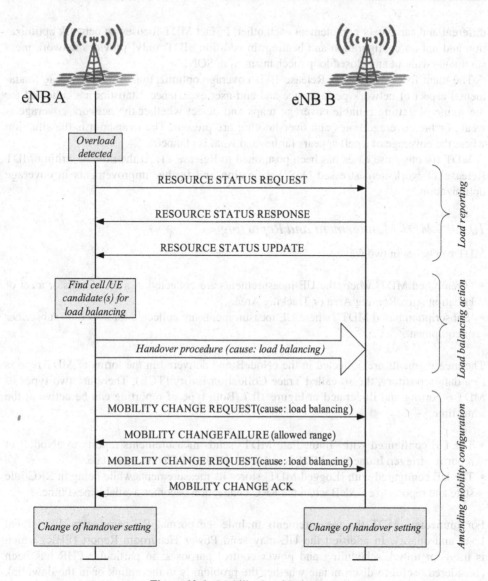

Figure 10.6   Mobility Load Balancing.

terminals, measuring devices and additional tools. The collected information is analysed off-line and used to understand what parameters need adjustments.

Because the current methodology is quite expensive, the automation of drive tests became one of the operators' top requirements in 3GPP. The target of MDT is to use commercial UEs with measurement logging plus location information. Commercial UEs naturally provide signal strength and quality when moving around the network and, having many UEs measuring, a continuous reporting is assured.

Although MDT and SON have the same high-level objectives; that is, to increase network performance and reduce operational efforts, their specific scope and actions are quite different and can nicely complement each other. In fact MDT focuses on network optimization and not on configuration and healing; in addition MDT 'only' provides network measurements without any closed loop mechanism as in SON.

The main focus of MDT in Release 10 is coverage optimization; coverage is one fundamental aspect of network performance and end-user experience. Statistics are collected for the scope of getting reliable coverage maps and detect whether the network coverage is weak, or the coverage holes and overshooting are present. The overshoot is the situation where the coverage of a cell appears far beyond what is planned.

MDT for other use cases has been postponed to Release 11. At the time of writing MDT Release 11 work item stressed QoS Verification and further improvements in coverage optimization.

10.3.4 MDT Management and Reporting

MDT can be set in two ways:

- 'Area based MDT' where the UE measurements are collected in a set of cells or a set of Location Area/Routing Area or Tracking Area.
- 'Subscription based MDT', where UE measurements are collected for a specific subscriber or equipment.

The measurements are collected in the eNodeB and delivered in the forms of MDT reports to a data repository, the so called Trace Collection Entity (TCE). There are two types of MDT reporting and illustrated in Figure 10.7. Both type of reporting can be active at the same time.

- The UE configured with 'Immediate MDT' sends measurements reports to eNodeB at reporting trigger. Immediate MDT is active for UE in RRC connected state.
- The UE configured with 'Logged MDT' stores its measurements while being in RRC idle state and reports to eNodeB when the RRC connection becomes available next time.

For Immediate MDT the measurements include the normal RF parameters, like signal level and quality. In addition the UE may send Power Headroom Report (PHR) which is used for uplink scheduling and power control purposes; in particular PHR has been considered useful to discriminate whether the problem is in the uplink or in the downlink

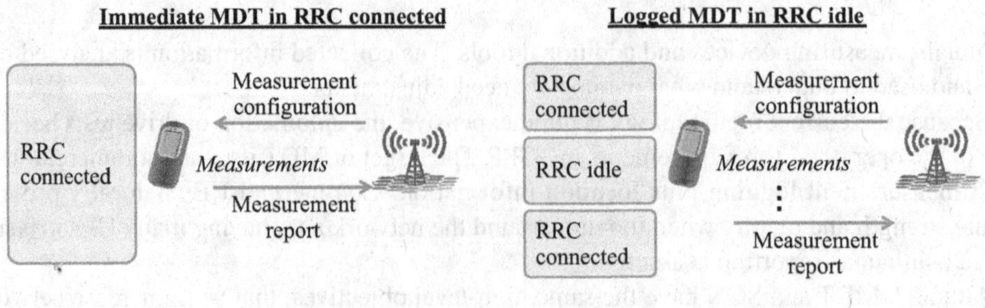

Figure 10.7 Immediate and logged MDT measurements.

connection. PHR is sent on MAC layer and it is up to the eNodeB to include it in the MDT report to the TCE.

In the Logged MDT, the UE logs those measurements available from normal cell reselection procedures, so that logged MDT does not require specific configuration of measurement parameters.

Uplink measurements are also considered for MDT; SINR and Uplink signal strength measured by eNodeB. The way those uplink measurements are used is vendor specific.

MDT measurements can include also the UE position information and time stamp. The UE location in Release 10 is provided in a best effort manner, which depends on terminal capability and network conditions. There are three ways for determining UE position:

- Cell ID (always known with Immediate reporting);
- RF fingerprint (using neighbours' cell measurements);
- Standalone GNSS positioning.

The measurement configurations need to consider the UE battery life. If the measurements are performed and reported too frequently, it may impact the UE power consumption.

10.3.5   Energy Savings

In the last few years, the energy consumption has become an important issue for mobile operators; aim is to minimize energy consumption while preserving the quality of service perceived by users.

Operators' energy consumption awareness is due to several reasons:

- Data booming forces the operators to install more base stations and new radio technologies;
- Increasing of energy costs;
- Kyoto-protocol: Operators might be forced to buy CO_2 certificates, if a pre-defined amount of energy consumption will be exceeded or there are even limits in some countries regarding maximum energy consumption of an operator per year.

The radio access network, and in particularly the base stations, have been identified to have the highest share of the mobile networks' overall energy consumption and hence the largest opportunity for energy saving measures. The networks are typically dimensioned to fulfil the higher traffic demands even if the off-peak periods would require much less capacity. A large potential for SON-based ES originates from the users' traffic profile. Usually night traffic profile is very different from the one in day-time; for example many users require voice calls during the day (e.g. in working hours) while the requests for data calls increase in late evening and night-time. Figure 10.8 shows an example of urban area and reflects the situation today; most likely in future there will be a shift towards more data traffic.

There are a number of ways to save energy.

- The most efficient one is the deployment and replacement of old equipment with state-of-the-art energy efficient hardware.

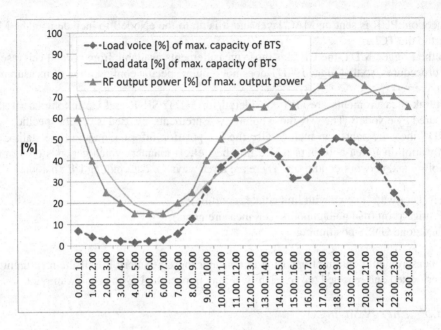

Figure 10.8 Example of traffic profile in a day.

- The type of power generator is a major energy consuming component; a great deal of energy can be saved by replacing the old diesel-based on site generator with on-site renewable (solar, wind, etc.) energy production.
- Vendors include own specific solutions, for example:
  - In case multiple frequency layers shut-down the second (and third) carrier.
  - Disable one MIMO path in case of LTE. Use lower transmission power during low traffic period but keeping same coverage.
- Switch off nodes with very low or not load at all. Base stations switching on and off is the solution indicated by 3GPP. Release 9 addressed the Intra-LTE case and defines basic X2 signalling for eNodeB communication, that is, switching off notification and switching on command. Reference [7] is a Release 10 Study Item and contains additional scenarios, for example Multi-RAT, and mechanisms (OAM based or through signalling across RATs without OAM intervention). Finally the Work Item was approved in Release 11.

Reference [7] addresses a number of scenarios: Inter-RAT, Inter-LTE and Intra-LTE. In the following those are split in two groups, the 'eNodeB Overlay' and the 'Capacity-Limited'.

10.3.6 eNodeB Overlay

Figure 10.9 includes a coverage layer and one or more capacity boosting cell(s) within its coverage. The coverage layer may be provided by LTE or a legacy RAT (2G/3G). Here a complete switch-off of the hotspot during low load/no load is no problem since the coverage is maintained by the macro cell. The cell switch on/off can be executed via OAM

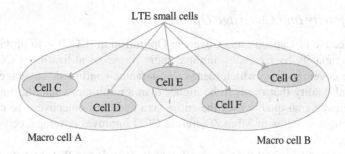

Figure 10.9 eNodeB overlay scenario.

mechanisms or via signalling across RATs without OAM involvement; the related neighbours shall be informed about the switch off decision and may request to switch on in case their load is getting too high. In fact due to the non-uniform users' distribution, the macro may risk to switch on cells which are of no help for offloading since users are located in a different area.

10.3.7 Capacity-Limited Network

Figure 10.10 shows the basic principle: during peak hours all cells are in no-ES mode since they are needed to manage the offered traffic; that is to guarantee the target grade of service. In off-peak periods the number of active cells may be lowered, and one given cell may compensate and ensure entire coverage. This scenario is a quite complicate one since it dynamically changes the cell density and switch on/off actions need to be accompanied by further modifications, as for example antenna tilting and transmit power. The scenario was not included into Release 11.

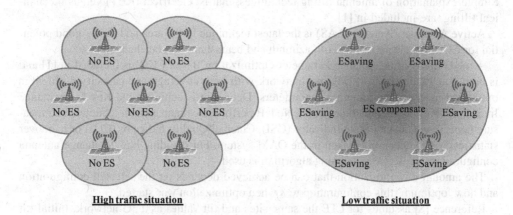

Figure 10.10 Capacity limited network. ES = Energy Savings.

10.3.8   Capacity and Coverage Optimization

The main objective of Capacity and Coverage Optimization (CCO) is to provide sufficient coverage and capacity by means of minimal radio resources utilization. CCO has first to maximize the coverage area, which means to guarantee continuous coverage in terms of received signal quality that needs to be higher than a minimum value. Secondly it has to provide sufficient signal quality over the entire area and third objective is to maximize the system throughput in terms of Mbps. Therefore CCO improves coverage, cell edge bit rate and cell throughput.

CCO is very expensive and time-consuming task, considering that the network coverage and capacity can vary due to changed environment. The environment is subject to continuous variations: examples of factors impacting the environment are new buildings' constructions and/or season changes such as trees with or without leaves, and many more. If the network is not regularly controlled coverage and capacity may be reduced since environment changes from initial assumptions.

Those environment changes are in nature very slow, and there is no need for a fast reaction. Due to the slow nature, the triggers could be Key Performance Indicator (KPI) values crossing related thresholds.

Coverage and capacity can be optimized in several ways, for example by adjusting antenna parameters. However this would require an antenna with Remote Electrical Tilt (RET) or Active-antennas in the base station. Appropriate signal level could be adjusted in different cell corners by means of adjusting antenna tilting, half power beam width or antenna direction. Alternatively downlink transmission power, transmitted reference signal power or power control parameters can be adjusted. 3GPP approach is for a centralized architecture (TS 32.522) and CCO logic is located in the OAM system.

RET is becoming more popular in nowadays since it removes the need for climbing towers and visit to base station sites and the tilt angle can be controlled remotely via OAM. However, the electrical tilt has some limitations and mechanical tilt is also needed. Although RET provides great advantages, it should carefully considered since changing antenna parameters, like antenna tilt, is not a trivial action because it affects the cell boundary and a wrong decision can lead to poor coverage or high interference. Simple explanation of antenna tilting techniques, that is, electrical (RET) versus mechanical tilting, are included in [1].

Active Antennas Systems (AAS) is the latest technique in the area and shows good potential for CCO. By means of AAS tilt, azimuth and beamshape can be changed.

A methodology for 'Antenna Parameter Optimization Based CCO' is included in [1] and is based on studies from [8]. Here a network with both coverage and capacity problem is optimized by changing antennas' parameters. Detection is achieved via Key Performance Indicators (KPIs) which includes both eNodeB performance management counters and measurements (e.g. Channel State Indicator (CSI), Channel Quality Indicator (CQI) or handover statistics); CCO decision is taken in the OAM system. For deciding how to change antenna configuration, a case based learning algorithm is used.

The amount of optimization that can be achieved depends on the selected configuration and how 'optimum' this configuration was when optimization was started.

Reference [8] assumes for LTE the same sites and tilt values as a 3G network. Initial tilt angles were planned according to 3G criteria, that is considering soft handover gains (while

LTE use hard handover). As study shows, after the self-optimization of remote electrical tilt the tilt angles are increased and cell overlapping area gets smaller. Performance results when optimizing antenna parameters are shown in Figure 10.11: the fiftieth percentile and tenth percentile of the SINR CDF represent the capacity and coverage improvements when changing antennas parameters; fiftieth percentile represents group of UEs, which indicate the average performance of the cell, and tenth percentile represents group of UEs with worst

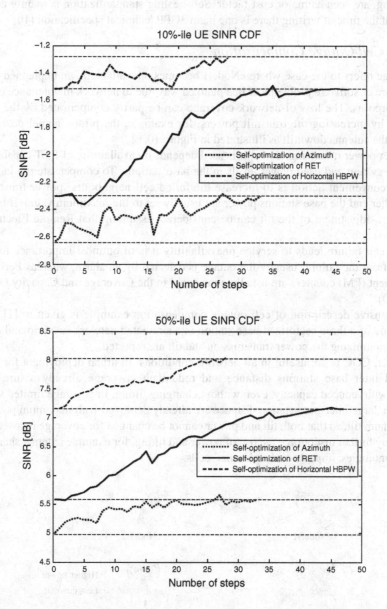

Figure 10.11 The tenth (upper) and fiftieth -percentile and points of SINR CDFs after each optimization step [9].

channel conditions and located near the cell edge. Figure 10.11 shows that the network capacity has more significant SINR gains than network coverage.

10.4 Self-Healing

Within the telecommunication industry, the operation of the network and in particular troubleshooting, are seen as major cost factor. Self-healing standardization is mainly addressed by SA5; at the time of writing there is one main 3GPP technical specification [10].

10.4.1 Cell Outage Compensation

Cell outage refers to the case where eNodeB becomes unavailable in an unplanned way due to hardware or software problems or in a planned way due to network maintenance or energy saving purposes. The loss of network coverage can be partly compensated by the adjacent eNodeBs by increasing the transmit powers, for example, the pilot channel power, or by reducing the antenna downtilt as illustrated in Figure 10.12.

Whether power or antenna downtilt is used depends on availability of RET-capable antennas in the system and used power levels in the base stations. To compensate the failed area, the most convenient action is to increase the failed-cell neighbours' power transmission. On the other end the base stations power are usually set to the maximum value. If this is the case, the re-adjustment of the tilt can be considered, assuming that Remote Electrical Tilt (RET) is available.

As the cell failure leads to service unavailability it is of outmost importance to recover very fast from the failure; usually the outage is detected by an alarm, whereas Performance Management (PM) counters are usually considered in the Coverage and Capacity Optimization (CCO).

An extensive description of cell outage compensation example is given in [1]. Here, a malfunction of a three-sectorized base station in a simulated network scenario and compensation by optimizing the power transmission and tilt are reported.

However, COC is not useful in all network conditions. In urban deployment for example the small inter base stations distance and redundant coverage already assure service, although with reduced capacity, even without changing tilting. In coverage limited scenarios instead the large inter-base station distance are already deployed with maximum power level and minimum tilt, so that both tilt and power cannot be changed for coverage recovery. Other factors may further discourage the use of power and tilting, for example antenna sharing with other technologies, impacts on neighbours' cells.

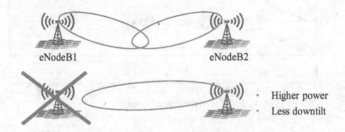

Figure 10.12   eNodeB1 outage compensation.

10.5   SON Features in 3GPP Release 11

3GPP Release 11 work started with following SON features:

- Mobility Robustness Optimization (MRO) enhancements. MRO was included in Release 10 for intra-LTE cases. Release 11 enhances the functionality to inter-RAT cases.
- Ping-pong and short-stay minimization in intra-RAT and inter-RAT cases. The target is to minimize unnecessary mobility events and the related signalling and UE power consumption.
- Optimal RAT selection based on QoS requirements and each RAT QoS capability.
- Enhancements to automatic neighbour relations to inter-RAT cases.
- Coordination between MRO and other traffic control mechanism, like mobility load balancing.
- Studying methods to verify the status of the cell's radio resources.

After the evaluation phase, at RAN3 #75 (February, 2012), the Release 11 Work Item was limited to the topic identified as the most relevant, that is those addressing failure scenarios and inter-RAT ping-pong.
   Failure scenarios:

- Failure while moving out of LTE coverage: connection fails in LTE, UE reconnects at 2G or 3G. The handover at LTE has not been triggered in time.
- Failure when entering LTE coverage: a handover is triggered toward LTE from 2G or 3G, but the LTE connection fails either during handover, or soon after and the UE reconnects to the original RAT. The handover to LTE has been triggered too early.
- In intra-LTE HetNet deployment: a handover to a pico cell is triggered too early or to a wrong target; the existing Release 9/10 mechanisms do not allow to verify which handover criteria are to be corrected (example differentiations: fast/slow UE, eICIC-capable UE, or older device, etc.).

Inter-RAT ping-pong detection:

- Detection of an inter-RAT ping-pong and enabling collection of cell-specific ping-pong statistics; if the UE does not return to the same cell that the one where the ping-pong started (e.g. when a HO is triggered from cell A in LTE to 3G and then the UE returns to LTE, but to a different cell, e.g. cell B, if cell A should be notified about the problem).

Regarding the energy saving, the focus of Release 11 is to enable inter-RAT mechanism to request a pico cell to wake up and notify neighbours that a cell is suspended. The Work item also addresses a mechanism to detect which pico should be awaken in a scenario where the macro cell is overloaded and several picos are under its coverage.

10.6   Summary

The first set of SON features was included into 3GPP Release 8. The target of SON is to improve the network performance, to minimize the operating expenses, to improve the network resource usage and to help minimizing the impact of cell outage. The first SON algorithms focused on self-configuration while the further 3GPP releases included more features on self-optimization and self-healing. Release 9 brought mobility optimization and energy savings while Release 10 included capacity and coverage optimization and minimization of drive testing. Minimization of drive testing takes benefit of commercial UEs to provide feedback to optimize the network – minimizing the need for costly drive testing. Further 3GPP work focuses especially on inter-RAT optimization between LTE and HSPA and on SON features for LTE-Advanced radio evolution.

References

1. Hämäläinen, S., Sanneck, H. and Sartori, C. (2011) *LTE Self-Organizing Networks (SON)*, John Wiley & Sons, Ltd, Chichester.
2. 3GPP TR3.023 (2010) Technical Specification Group Radio Access Network, Evolved Universal Terrestrial Radio Access Network (E-UTRAN); Self-configuring and self-optimizing network (SON) use cases and solutions, ver.0.1.0., Release 10, 24 June.
3. 3GPP R3-100916 (2010) SON_replyLS_RAN2_disc, 2010.
4. R3-102341 (2010) SON_MRO_idle_req: Enabling UE-originated RLF reporting in case of RRC connection setup, Nokia Siemens Networks, August.
5. R3-102342 (2010) SON_MRO_idle_info: Information to be reported in the UE-originated RLF reporting in case of RRC connection setup, Nokia Siemens Networks, August.
6. 3GPP TS 36.413 (2011) Technical Specification Group Radio Access Network; Evolved Universal Terrestrial Radio Access Network (E-UTRAN); S1 Application Protocol (S1AP) (Release 10).
7. 3GPP TR36.927 (2011) Technical Specification Group Radio Access Network, Potential Solutions for Energy Saving for E-UTRAN, ver.10.0.0., Release 10, 24 June.
8. Yilmaz, O.N.C., Hamalainen, S. and Hamalainen, J. (2009a) Analysis of antenna parameter optimisation space for 3GPP LTE. Proceedings of IEEE Vehicular Technology Conference (VTC), Anchorage, Alaska, September.
9. Yilmaz, O.N.C., Hamalainen, S. and Hamalainen, J. (2009b) Comparison of remote electrical and mechanical antenna downtilt performance for 3GPP LTE. Proceedings of IEEE Vehicular Technology Conference (VTC), Anchorage, Alaska, September.
10. 3GPP TS32.541 (2011) Technical Specification Group Services and System Aspects; Telecommunication management; Self-Organizing Networks (SON); Self-healing concepts and requirements.

11

Performance Evaluation

Harri Holma and Klaus Pedersen

11.1 Introduction

The International Telecommunication Union (ITU) defined requirements for the new International Mobile Telecommunications (IMT)-Advanced systems. The target was that the new systems can provide substantially higher performance than those IMT-2000 systems 10 years earlier. The targets were defined for the peak and average data rates and for the spectral efficiency in multiple environments, for latency, for mobility and for spectrum flexibility. Additionally, 3GPP defined its own targets for LTE-Advanced. 3GPP targets were set higher than those defined by ITU. This chapter presents the target setting and the results of the performance evaluation.

LTE-Advanced enhances system performance compared to LTE Release 8 in multiple operating domains: in spatial domain, in time domain and in frequency domain. The main enhancements are listed in Figure 11.1. The spatial domain refers to new capabilities to take benefit of multi-antenna solutions for increasing the data rates and for increasing the spectral efficiency with beamforming. The multi-antenna techniques are enhanced also to multi-cell cases with Coordinated Multipoint (CoMP) transmission and reception. LTE-Advanced allows utilizing heterogeneous networks including small cells and relay nodes. The frequency domain is enhanced by carrier aggregation up to 100 MHz bandwidth and by uplink multi-cluster scheduling. Those new features can improve LTE-Advanced performance compared to Release 8.

This chapter presents the performance targets defined by ITU and by 3GPP in Section 11.2 and the results of the performance evaluation is shown in Section 11.3. The maximum capacity per subscriber with LTE-Advanced features is estimated in Section 11.4 together with the coverage discussions. The chapter is summarized in Section 11.5.

LTE-Advanced: 3GPP Solution for IMT-Advanced, First Edition. Edited by Harri Holma and Antti Toskala.
© 2012 John Wiley & Sons, Ltd. Published 2012 by John Wiley & Sons, Ltd.

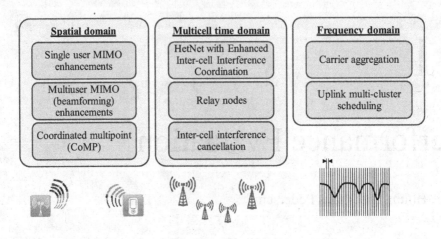

Figure 11.1 Summary of LTE-Advanced performance solutions.

11.2 LTE-Advanced Targets

ITU set performance targets for IMT-Advanced in [1]. IMT-Advanced systems need to support very high data rates in local area, and high mobility and flexibility for multiple services. The peak spectral efficiency in downlink with 4×4 MIMO must be 15 bps/Hz and in uplink with 2×4 MIMO 6.75 bps/Hz. The bandwidth must be scalable up to 40 MHz and preferably up to 100 MHz. The peak data rate would be 1.5 Gbps in downlink and 675 Mbps in uplink with 100 MHz. ITU also defined the latency requirements: control plane latency from idle to active must be below 100 ms and one way user plane latency below 10 ms. The handover break must be less than 27.5 ms for intra-frequency handover and below 60 ms for inter-band handover.

The cell spectral efficiency target is defined both for average case and for the cell edge case. The cell edge is defined as the 5% point of the cumulative distribution function of the user throughputs. The capacity for Voice over IP (VoIP) is defined per MHz with 50% voice activity and with the maximum one way radio access delay of 50 ms. VoIP capacity is the minimum value of uplink and downlink capacity. The antenna configuration assumes two antennas in the terminal and four antennas at base station, downlink 4×2 and uplink 2×4. ITU requirements for average and cell edge efficiency in downlink are shown in Table 11.1

Table 11.1 ITU and 3GPP requirements for IMT-Advanced for downlink with 4×2 antennas [bps/Hz/cell]

	Average spectral efficiency	Cell edge efficiency	VoIP [users/MHz/cell]
Indoor	3.0	0.100	50
Microcells	2.6	0.075	40
Base coverage urban	2.2	0.060	40
High speed rural	1.1	0.040	30
3GPP target Case 1	2.6	0.090	not specified

Table 11.2   ITU and 3GPP requirements for IMT-Advanced for uplink with 2 × 4 antennas [bps/Hz/cell]

	Average spectral efficiency	Cell edge efficiency	VoIP [users/MHz/cell]
Indoor	2.25	0.070	50
Microcells	1.8	0.050	40
Base coverage urban	1.4	0.030	40
High speed rural	0.7	0.015	30
3GPP target Case 1	2.0	0.070	not specified

and for uplink in Table 11.2. Four different environments are used: indoor, microcells, urban macro coverage and high speed rural mobility [2]. The highest requirements are defined for indoor case due to favourable propagation condition and second highest in microcells while the requirements are less demanding in urban macrocells and in high mobility cases. In general, the downlink requirements are 30–60% higher than the uplink requirements. The reason is that the point-to-multipoint scheduling in downlink is more efficient than multipoint-to-point in uplink. The downlink also has more transmission power available. The cell edge requirements are approximately 3% of the average requirements. The cell edge data rates are mainly limited by the inter-cell interference.

3GPP defined its own set of targets for LTE-Advanced in Reference [3]. 3GPP targets were higher than ITU requirement. 3GPP also defined the targets for different antenna configurations with two and four base station antennas, two and four terminal receiver antennas, and one and two terminal transmit antennas. We mainly focus on 3GPP Macro Case 1 which is similar to ITU base urban coverage macrocell.

11.2.1   ITU Evaluation Environments

The indoor case emulates isolated cells in public premises or in the office environments. The users are stationary or moving at pedestrian speeds. The scenario consists of one floor with 16 rooms of 15 × 15 m and a long hall of 120 × 20 m. Two base stations are placed in the middle of the hall at 30 m and 90 m locations. The indoor scenario is shown in Figure 11.2. The base station power is 24 dBm and the antenna gain is 0 dBi. The indoor case uses 20 MHz bandwidth while other scenarios use 10 MHz. The number of users is 10 per cell, assuming a full buffer traffic model.

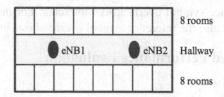

Figure 11.2   Indoor environment.

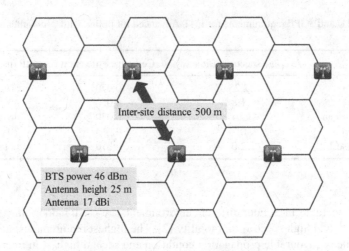

Figure 11.3 Base urban coverage with macrocells.

The microcell environment models high capacity dense urban solution with 200 m inter-site distance and 41 dBm base station power. The base station antenna is located at 10 m height with maximum gain of 17 dBi. The deployment scenario is the Manhattan grid and some of the users have line-of-sight connection. The microcellular test environment includes outdoor and outdoor-to-indoor users: in the latter case the users are located indoors and base stations outdoors. Therefore the channel model for the microcellular test environment contains two parts, the outdoor part and the outdoor-to-indoor part. The line-of-sight probability equals 60% for microcellular case.

The base urban coverage case has higher inter-site distance (500 m), higher antenna height (25 m) above rooftops and higher base station power of 46 dBm than microcell environment. The environment targets for continuous coverage for pedestrians and up to fast vehicular users. The line-of-sight probability equals 28% for urban coverage case. For non-line-of-sight links, the average azimuthal dispersion at the base station equals 26 degrees. The base urban coverage scenario is described in Figure 11.3.

The high speed environment models vehicles and trains with 120 kmph in large cells with inter-site distance of 1732 m and base station antenna height of 35 m. The base station antenna front-to-back ratio equals 20 dB for both the micro-, base coverage urban, and high speed rural environment. The effect of antenna down-tilt is included for those three environments as well, while the simple two-dimensional antenna modelling is assumed for the indoor case.

The overhead from control channels and reference symbols is taken into account in the simulations. For the downlink, the first three symbols per TTI are reserved for control channels, and also the reference symbol overhead is subtracted. For the uplink, four PRBs are reserved for PUCCH, and also the reference symbol overhead is subtracted.

11.3 LTE-Advanced Performance Evaluation

11.3.1 Peak Data Rates

The peak data rate increases with wider bandwidth and with more antennas. The LTE-Advanced peak data rates with different multiantenna configurations are illustrated in

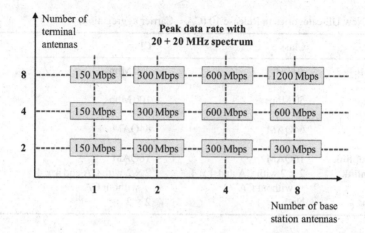

Figure 11.4 Downlink peak data rate with different antenna configuration with 20 + 20 MHz.

Figure 11.4. The minimum UE requirement is two receiver antennas. The calculation assumes carrier aggregation of 20 + 20 MHz and 64QAM modulation. The peak rate with single stream is 150 Mbps, dual stream 300 Mbps, four streams 600 Mbps, and with the maximum 8 × 8 MIMO beyond 1 Gbps. If 100 MHz spectrum would be used, the LTE-Advanced peak rate would be 3 Gbps. LTE-Advanced gives peak efficiency of 30 bps/Hz with 8 × 8 MIMO and 15 bps/Hz with 4 × 4 MIMO which exceeds with ITU performance requirements. ITU evaluation environments are summarized in Table 11.3.

11.3.2   UE Categories

3GPP Release 8 defined five UE categories with typical commercial devices using Class 3 and 4 with downlink peak rate of 100–150 Mbps with 2 × 2 MIMO and uplink peak rate of 50 Mbps. Release 10 adds three new UE categories shown in Table 11.4. Class 6 and 7 push

Table 11.3 ITU evaluation environments

	Indoor	Micro	Base coverage urban	High speed rural
Inter-site distance	60 m	200 m	500 m	1732 m
Carrier frequency	3.4 GHz	2.5 GHz	2.0 GHz	800 MHz
Base station antenna height	6 m	10 m	25 m	35 m
Base station power	21 dBm	41 dBm	46 dBm	46 dBm
Base station antenna gain	0 dBi	17 dBi	17 dBi	17 dBi
Terminal power	21 dBm	24 dBm	24 dBm	24 dBm
Terminal speed	3 kmph	3 kmph	30 kmph	120 kmph
Bandwidth (FDD case)	20 MHz	10 MHz	10 MHz	10 MHz
Number of users per cell	10	10	10	10
Service	Full buffer	Full buffer	Full buffer	Full buffer

Table 11.4 New UE categories in Release 10 (CA = Carrier aggregation)

UE category	Class 6	Class 7	Class 8
Peak data rate downlink	300 Mbps	300 Mbps	3000 Mbps
Peak data rate uplink	50 Mbps	100 Mbps	1500 Mbps
Modulation downlink	64QAM	64QAM	64QAM
Modulation uplink	16QAM	16QAM	64QAM
MIMO downlink	2×2 with CA and 4×4 without CA	2×2 with CA and 4×4 without CA	8×8
MIMO uplink	No	2×2	4×4

the downlink peak rate to 300 Mbps by using either 4×4 MIMO with 20 MHz or by 2×2 MIMO combined with $20 + 20$ MHz carrier aggregation. The uplink capability in Class 7 is increased to 100 Mbps by using 2×2 MIMO. Class 8 is the marketing category in 3GPP to show the very high peak rates: 3 Gbps downlink and 1.5 Gbps uplink by combining all the tricks of Release 8: 5×20 MHz of carrier aggregation, downlink MIMO 8×8, uplink MIMO 4×4 and uplink modulation of 64QAM.

11.3.3 ITU Efficiency Evaluation

This section presents the performance evaluation against ITU targets. LTE-Advanced average and cell edge spectral efficiency was evaluated in system simulations by multiple 3GPP companies. The detailed results are presented in [4]. The average spectral efficiency is presented in Figure 11.5 and the cell edge efficiency in Figure 11.6. CoMP and heterogeneous networks features are not included in the simulations. We can note a few aspects from these results:

- The simulated performance exceeds ITU targets in all cases. The difference is at least 25% except for the uplink average efficiency in micro and urban base coverage cases where the difference is below 10%.
- The cell edge efficiency in downlink is 3% of the average efficiency and just 1% of the peak data rate. That clearly illustrates the challenge in providing high consistent data rates at the cell edge in the presence of high inter-cell interference. The cell edge efficiency in uplink is 5% of the average efficiency and 1% of the peak data rate.
- The average efficiency in downlink is 75% higher than the uplink efficiency due to point-to-multipoint transmission and higher power. The difference in the cell edge efficiency between downlink and uplink is smaller – just 20%. The downlink cell edge users are impacted more than uplink users by the inter-cell interference.
- The indoor environment provides substantially higher capacity than other environments due to less inter-cell interference. The indoor capacity is nearly double compared to microcells and more than double compared to urban base coverage.
- The microcell average efficiency is +25% higher than urban base coverage case.

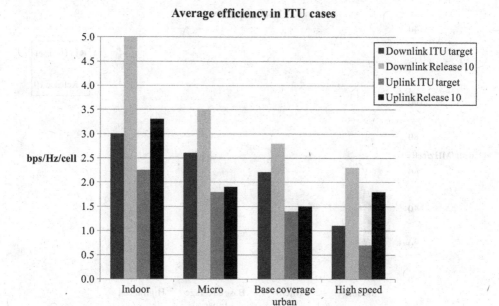

Figure 11.5   Average spectral efficiency in different ITU environments.

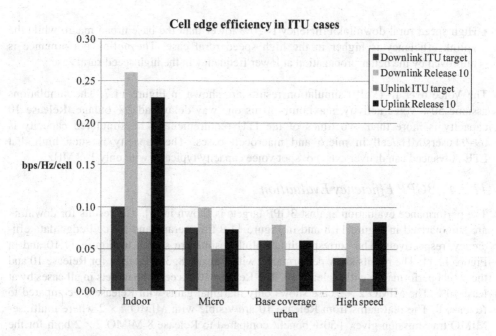

Figure 11.6   Cell edge spectral efficiency in different ITU environments.

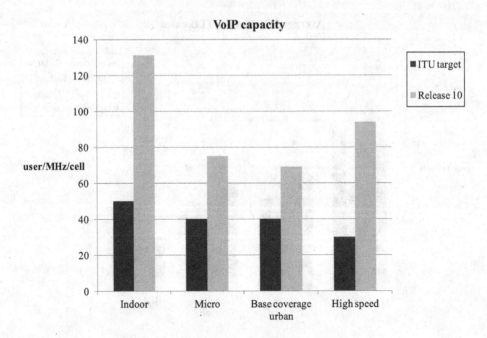

Figure 11.7 VoIP efficiency in different ITU environments.

• High speed rural downlink efficiency is 20% lower than the base urban macro while the uplink efficiency is higher in the high speed rural case. The uplink performance is improved by the better propagation at lower frequency in the high speed rural case.

The Voice over IP (VoIP) simulation results are shown in Figure 11.7. The simulations assume 50% voice activity, maximum 50 ms one way delay and 2% outage. Release 10 capacity is more than two times of the ITU requirements. The simulated capacity is 69–94 users/MHz/cell in micro and macrocell cases. The capacity is such high that LTE-Advanced can deliver even hot spot voice capacity typically with only 1–2 MHz.

11.3.4   3GPP Efficiency Evaluation

The performance evaluation against 3GPP targets is shown in [5]. The results for downlink are summarized in Figure 11.8 and in Figure 11.9 for average and for cell edge data efficiency, respectively. The corresponding uplink results are shown in Figure 11.10 and in Figure 11.11. The results show performance with Release 8, 3GPP target for Release 10 and the actual performance with Release 10. The Release 10 exceeds the targets in all cases by at least 10%. The MIMO 2×2 case shows only marginal gains with Release 10 compared to Release 8. The real gains from Release 10 are visible with MIMO 4×2 where multiuser MIMO transmission gives +50% benefit compared to Release 8 MIMO 2×2 both for the average and for the cell edge cases. Even higher efficiency can be achieved by having four antennas in UE: MIMO 4×4 gives 100% gain in average efficiency and up to +150%. The

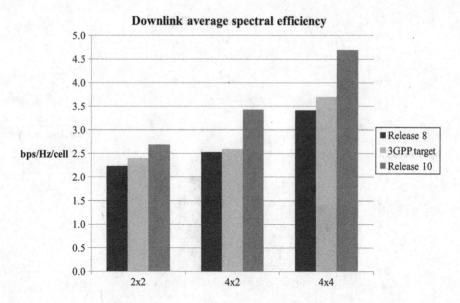

Figure 11.8   Downlink average spectral efficiency in 3GPP Case 1.

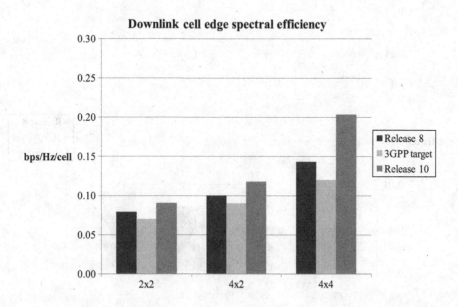

Figure 11.9   Downlink cell edge spectral efficiency in 3GPP Case 1.

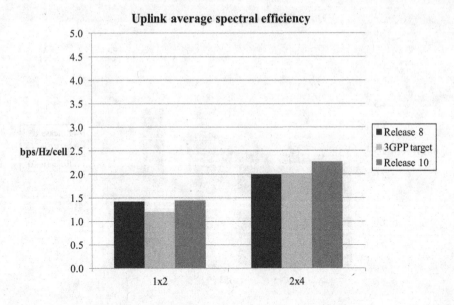

Figure 11.10   Uplink average spectral efficiency in 3GPP Case 1.

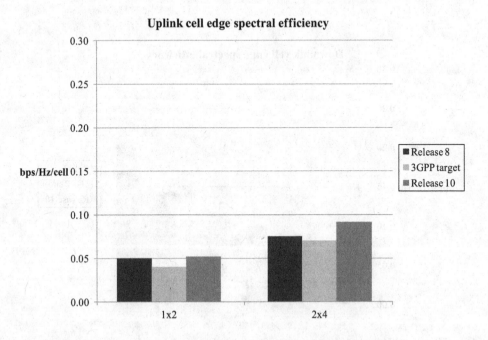

Figure 11.11   Uplink cell edge spectral efficiency in 3GPP Case 1.

average efficiency with MIMO 4×4 is 4.7 bps/Hz/cell, which is close to 100 Mbps in 20 MHz channel or 500 Mbps in 100 MHz channel.

The uplink efficiency improves considerably by having four antennas at base station compared to two antennas. The four antenna benefit is 60–80%. Most of the gain can be obtained with Release 8 as well while Release 10 provides 10–20% additional benefits.

11.4   Network Capacity and Coverage

The new devices have increased the data usage considerably. The typical data usage values per subscriber per month are shown in Figure 11.12. The larger the screen size, the higher the data usage. The smartphone users consume 300 to 500 Megabyte (MB) of data per month, tablet users consume more data and laptop users consume normally several GB of data per month. The data usage numbers also tend to increase continuously due to new applications, higher access speeds and better quality displays.

The following calculation provides an estimation of the maximum radio capacity that can be provided with LTE and LTE-Advanced radios in typical network configurations. The assumptions are listed in the following:

- The total spectrum for LTE per operator is assumed 2×50 MHz (FDD). The typical frequencies in Europe would be at 800, 1800 and 2600 MHz bands. Another scenario assumes two times more spectrum: 2×100 MHz per operator including more spectrum from the 3.5 GHz band.
- Only downlink traffic is considered because typical traffic is 5–10 times more in downlink than in uplink. We assume that the network capacity is downlink limited.
- 10.000 base station each with three sectors.
- 20 million subscribers.
- LTE assumes 2×2 MIMO with spectral efficiency of 2.7 bps/Hz/cell and LTE-Advanced 4×4 MIMO with 4.7 bps/Hz/cell.
- Macrocell sectorization brings additional 50% more efficiency.
- Small cells with heterogeneous networks provides 250% more capacity.
- Geographical traffic distribution with 15% of the sectors carrying 50% of the traffic.
- Traffic distribution in a 24 hour period: busy hour carries 7% of daily traffic.

Figure 11.13 illustrates the maximum capacity per subscriber per month in five different scenarios. The LTE deployment with 2×50 MHz spectrum enables 5 GB traffic which is

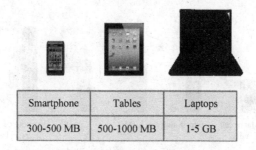

Smartphone	Tables	Laptops
300-500 MB	500-1000 MB	1-5 GB

Figure 11.12   Typical data usage per month per subscriber with different type of devices.

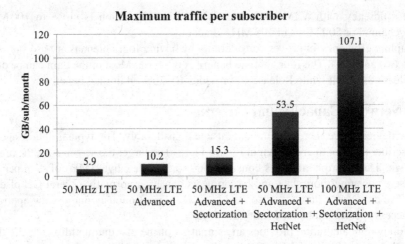

Figure 11.13   Radio capacity per subscriber per month.

enough for the current usage patterns. LTE-Advanced can squeeze out 10–15 GB from the existing macrocellular network. If the operator additionally deploys small cells in the busy areas, the capacity grows beyond 50 GB. If the operator can get more spectrum, the capacity will grow beyond 100 GB per subscriber per month. These capacities are approximately 100 times more than the current usages.

The network coverage is another challenge for providing higher data rates. It is not possible to increase the transmission power levels from terminals or from the base stations. It is also not possible to improve the receiver sensitivity because the noise figures are already close to the theoretical values. The main solutions for improving the coverage will be more base stations, more antennas and multi-cell reception. We assume that the initial LTE macronetwork is designed to provide 1 Mbps downlink and 100 kbps uplink at the cell edge (see Figure 11.14). The sectorization is estimated to improve the cell edge data rates by +50%

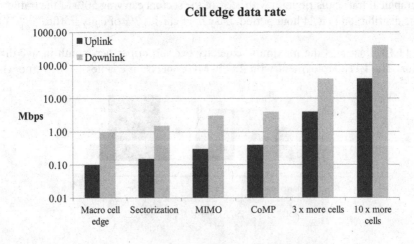

Figure 11.14   Radio coverage improvements.

with higher antenna gain, four-antenna MIMO by +100%, Coordinated Multipoint (CoMP) transmission and reception by +30%. These solutions together can improve the cell edge data rate by a factor of four. The major boost for the data rates can be obtained with more cells. If we have three times more cells, the cell edge link budget can be improved by 10 dB, which increases noise limited coverage by a factor of 10. If there are ten times more co-channel cells, the improvement can be a factor of hundred pushing the downlink cell edge data rates clearly beyond 100 Mbps and uplink to several tens of Mbps.

11.5 Summary

ITU defined performance targets for IMT-Advanced technologies in terms of peak data rates, average efficiency and cell edge efficiency in multiple different propagation environments. Also 3GPP defined its own performance targets for LTE-Advanced. This chapter illustrated the targets and the corresponding simulation results in 3GPP. LTE-Advanced can improve performance because of new features in spatial domain, time domain and frequency domain which enhance the system performance in case of multiantennas, heterogeneous networks and wideband systems. The result of the system simulations is that LTE-Advanced in Release 10 exceeds all ITU and 3GPP performance requirements. LTE-Advanced spectral efficiency is improved compared to Release 8 especially in multiantenna cases with 4×2 and 4×4 MIMO transmission. The macro cell spectral efficiency is up to 4.7 bps/Hz/cell with 4×4 MIMO corresponding to nearly 200 Mbps average efficiency with 40 MHz bandwidth.

The network level dimensioning shows that typical operator spectrum allocations with existing base station density and with LTE-Advanced efficiency can provide over 10 GB data for every subscriber per month. By macrocell sectorization, heterogeneous networks and new spectrum allocations enable even more than 100 GB data per month. In many cases the data rates are limited by the coverage, which likely requires new base stations to deliver the required link budgets.

References

1. Report ITU-R M.2134 (2008) Requirements related to technical performance for IMT-Advanced radio interface(s).
2. Report ITU-R M.2135-1 (2009) Guidelines for evaluation of radio interface technologies for IMT-Advanced.
3. 3GPP TR 36.913 V8.0.1 (2009-03) Technical Report, Requirements for further advancements for Evolved Universal Terrestrial Radio Access (E-UTRA) (LTE-Advanced).
4. 3GPP TR 36.912 V10.0.0 (2011-03) Technical Report, Feasibility study for Further Advancements for E-UTRA (LTE-Advanced).
5. 3GPP TR 36.814 V9.0.0 (2010-03) Technical Report, Further advancements for E-UTRA physical layer aspects.

12

Release 11 and Outlook Towards Release 12

Timo Lunttila, Rapeepat Ratasuk, Jun Tan, Amitava Ghosh and Antti Toskala

12.1 Introduction

This chapter presents the outlook towards 3GPP Release 11 and 12 with the work on-going in 3GPP. First the topics that were started and ongoing in 3GPP towards LTE-Advanced in Release 11 are introduced. Of the 3GPP Release 11 LTE-Advanced topics other items are covered in this chapter, with the exception of the Co-ordinated Multipoint Operation (CoMP) which is treated separately in Chapter 13. The Release 11 topics covered in this chapter are the advanced LTE UE receiver, carrier aggregation enhancements, low cost Machine Type Communications (MTC), and downlink MIMO enhancements with focus on control channel improvements. This chapter concludes with the outlook towards Release 12.

12.2 Release 11 LTE-Advanced Content

3GPP is complementing the first LTE-Advanced version, Release 10, with large number of new study and work items in Release 11. Perhaps the biggest topic currently on-going is the CoMP which was a part of the original LTE-Advanced studies but was eventually postponed to Release 11 to ensure finalization of the items covered in the previous chapters. The CoMP principles and performance are covered in detail in Chapter 13. The key 3GPP work items for Release 11 are shown in Figure 12.1, with the focus on the topics aiming to improve performance metrics such as network capacity. Respectively, the key study items are shown in Figure 12.2. There are likely to be further study items introduced during the Release 11 work but those will most likely end up finalizing the studies so that the actual features included in the specifications are only in later Releases. The full 3GPP TSG RAN Release 11 work program contains more topics than shown in Figure 12.1, but some of them present a continuation of LTE Release 8 and 9 topics, and thus are not considered to be

LTE-Advanced: 3GPP Solution for IMT-Advanced, First Edition. Edited by Harri Holma and Antti Toskala.
© 2012 John Wiley & Sons, Ltd. Published 2012 by John Wiley & Sons, Ltd.

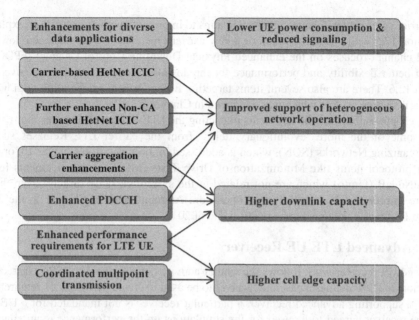

Figure 12.1   Release 11 key LTE-Advanced work items.

LTE-Advanced related features. The topic which received most interest in 3GPP in Release 11 prioritization activity from 3GPP network operators was the enhanced performance requirements for LTE UE, improving the system performance with advanced receivers as discussed in Section 12.3. From the listed topics the low cost MTC UE is not aiming to directly to increase capacity or data rates, but to facilitate LTE introduction to the mass market for the machine to machine communications by investigating how to ensure cost competitive radio modem offering, as is discussed more in Section 12.4. The enhancements for diverse data applications target low power consumption with the 'always on' type of LTE smartphone applications while minimizing the resulting signaling to the network as well.

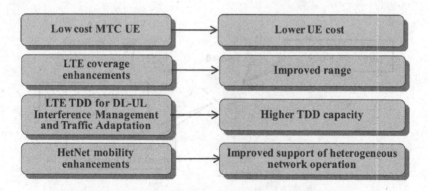

Figure 12.2   Release 11 key LTE-Advanced study items.

The carrier aggregation enhancements topic aims to improve downlink capacity as explained in Section 12.5 with the introduction of a new carrier type. The WI on enhanced downlink control channel focuses on the enhanced Physical Downlink Control Channel (ePDCCH) to offer better flexibility and performance for the downlink control signaling as covered in Section 12.6. There are also several items targeting further improved operation with heterogeneous networks, a topic which was addressed in Chapter 8.

This chapter introduces the key items impacting the LTE-Advanced features, but will not cover some of the more 'evolutionary items' from the earlier LTE Releases such as Self-Organizing Networks (SON), which is addressed in more detail in Chapter 10, or other similar protocol items like Minimization of Drive Tests (MDT) or enhancements for the home eNodeB (femto) which are not related to the system capacity directly even though they are important for overall system operation. As mentioned in Chapter 2, the 3GPP Release 11 items are planned to be ready at end of 2012.

12.3   Advanced LTE UE Receiver

Advanced UE receiver work follows the similar activity done earlier for HSPA UE receiver. Even though 3GPP defines a baseline receiver to be used to create performance requirements for a UE supporting advanced receiver, a particular receiver is not mandated for a UE. The baseline receiver agreed to be used for the simulations on the performance requirements is Minimum Mean Square Error (MMSE) – Interference Rejection Combining (IRC) receiver. The MMSE-IRC receiver has the capability to 'reject' the interference by creating a 'null' in the spatial domain towards the most dominant interferer, as was also discussed in [1]. Such an operation is illustrated in Figure 12.3.

Figure 12.4 shows the basic block diagram of an Advanced UE receiver for decoding PDSCH using Maximal Ratio Combining (MRC) and IRC principles. The received signal is down-converted to baseband in each antenna branch. For each symbol the Cyclic Prefix (CP) is first removed and the Fast Fourier transform (FFT) is performed on the block of samples. A separate channel estimation block performs the channel estimation based on common or dedicated reference symbols. MMSE-MRC or MMSE-IRC based algorithm equalizes the received symbols with the complex channel estimate. The equalized symbols are fed into the

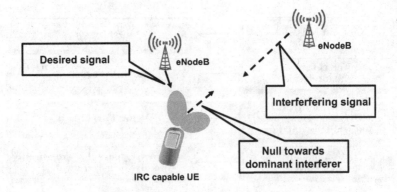

Figure 12.3   The MMSE-IRC receiver principle.

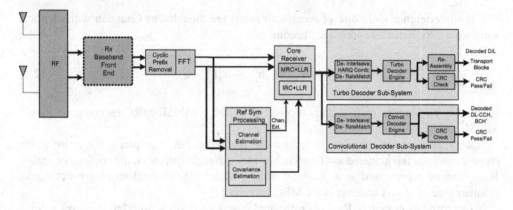

Figure 12.4 Basic block diagram of an UE receiver.

Log-Likelihood Ratio (LLR) calculation based on modulation types. The LLR symbols are de-interleaved and passed through the Turbo decoder engine to get decoded symbols.

12.3.1   Overview of MMSE-MRC and MMSE-IRC Methods

Let $x_{i,j,m}$ represent the FFT output sample corresponding to the ith OFDM symbol ($1 \leq i \leq 14$), jth sub-carrier ($1 \leq j \leq N$, where N is the number of sub-carriers allocated to the user of interest), and the mth receive antenna. The received signal can be modeled as

$$\mathbf{x}_{i,j} = \mathbf{h}_{i,j}\mathbf{s}_{i,j} + \mathbf{v}_{i,j} \tag{12.1}$$

where $\mathbf{x}_{i,j} = [x_{i,j,1}, x_{i,j,2}, \ldots, x_{i,j,M}]^T$, $\mathbf{h}_{i,j} = (h_{i,j,m,k})_{M \times K}$, $h_{i,j,m,k}$ is the effective channel gain between the k-th transmit antenna and the m-th receive antenna of the i-th symbol at the j-th sub-carrier, $\mathbf{s}_{i,j} = [s_{i,j,1}, s_{i,j,2}, \ldots, s_{i,j,K}]^T s_{i,j,k}$ is the complex modulation symbol transmitted on the jth sub-carrier of the ith OFDMA symbol at the kth transmit antenna, $\mathbf{v}_{i,j} = [v_{i,j,1}, v_{i,j,2}, \ldots, v_{i,j,M}]^T$, and $v_{i,j,m}$ is the signal component corresponding to noise and total interference. It is assumed that there are K number of transmit antenna and M number of receive antenna elements.

Using the minimum mean square error (MMSE) criteria, the expression for the estimated symbol becomes

$$\hat{\mathbf{s}}_{i,j} = \mathbf{h}_{i,j}^{H}(\mathbf{h}_{i,j}\mathbf{h}_{i,j}^{H} + \mathbf{R}_{vv,i,j})^{-1}\mathbf{x}_{i,j} \tag{12.2}$$

The correlation matrix $\mathbf{R}_{vv,i,j}$ is the spatial correlation of received interference and noise of M receive antennas and is given by

$$\mathbf{R}_{vv,i,j} = E\left[\mathbf{v}_{i,j}\mathbf{v}_{i,j}^{H}\right] \tag{12.3}$$

where E[.] denotes expectation.

If the interference and noise of receive antennas are modeled as Gaussian with identical variance σ_n^2, the detected symbol $\hat{s}_{i,j}$ becomes

$$\hat{s}_{i,j} = \mathbf{h}_{i,j}^H \left(\mathbf{h}_{i,j} \mathbf{h}_{i,j}^H + \sigma_n^2 \mathbf{I} \right)^{-1} \mathbf{x}_{i,j} \tag{12.4}$$

where $\mathbf{I}$ is an identity matrix. This receiver is called a MMSE-MRC receiver, or simply MMSE receiver [2].

Note, that there are other possible baseline receivers. For example, if the noise at the receive antennas are modeled as Gaussian but with different variances, the covariance matrix $\mathbf{R}_{vv,i,j}$ can be represented as a diagonal matrix where the diagonal terms are not equal. Another type of rank-1 receiver is the MRC receiver [2].

If the correlation matrix $\mathbf{R}_{vv,i,j}$ is estimated based on realistic interference over receive antennas, the corresponding MMSE receiver is called MMSE-IRC receiver, also known as advanced LTE UE receiver.

To estimate the interference-plus-noise correlation matrix, the vector $\mathbf{e}_{i,j} = [e_{i,j,1}, e_{i,j,2}, \ldots, e_{i,j,M}]^T$ of interference-plus-noise samples is computed from the reference signal symbols:

$$\mathbf{e}_{i,j} = \tilde{\mathbf{x}}_{i,j} - \hat{\mathbf{h}}_{i,j} \mathbf{d}_{i,j}, \quad (i,j) \in \Psi_P \tag{12.9}$$

where $\Psi_P \subset \{1, \cdots, 14\} \times \{1, \cdots, N\}$ is the set of reference signal symbol indices in sub-frames and sub-carrier, $\tilde{\mathbf{x}}_{i,j} = [\tilde{x}_{i,j,1}, \tilde{x}_{i,j,2}, \ldots, \tilde{x}_{i,j,M}]^T$ are received symbols at RS locations and $\hat{\mathbf{h}}_{i,j}$ is the estimate of $\mathbf{h}_{i,j}$ when non-ideal channel estimation is employed.

The reference symbol $\mathbf{d}_{i,j}$ is a length-K vector at sub-frame-i and sub-carrier-j. Note that for common reference signal (CRS)-based transmission, there is no precoding for the RS resource elements, while the precoding matrix is applied to the data symbols. The precoding matrix shall be applied in the channel estimation to yield the proper interference-plus-noise samples. For dedicated reference signal (DM-RS) based transmission, reference signals (RS) are precoded similar to their associated data symbols.

The estimated interference-plus-noise correlation matrix is then computed as

$$\hat{\mathbf{R}}_{vv} = \frac{1}{|\Psi_P|} \sum_{(i,j) \in \Psi_P} \mathbf{e}_{i,j} \mathbf{e}_{i,j}^H \tag{12.10}$$

where $|\Psi_P|$ is the number of RS symbols.

12.3.2 Performance of UE Receiver using IRC and its Comparison to MRC Receiver for Various DL Transmit Modes

In this section the link performances of MMSE-MRC versus MMSE-IRC are compared for downlink transmit mode 6 and 9. Due to the limited number of received branches of a LTE UE, rank-1 transmission is used to evaluate IRC performance. For CRS-based transmission, TM6 (or TM4 rank-1) is assumed; for DM-RS based, TM9 rank-1 is used. The bandwidth is selected as 10 MHz to test the MMSE-IRC algorithm throughput gain. A summary of

Table 12.1 Simulation assumptions for advanced receiver performance

Parameter	Scenario 1 (CRS based)	Scenario 2 (DM-RS based)
Carrier frequency	2 GHz	
System bandwidth	10 MHz	
Transmission mode on Serving cell	TM6	TM9 with 1-layer transmission
Transmission mode on interfering cell	TM6	TM9
MIMO configuration	2 × 2 and low correlation	4 × 2 and low correlation
Channel model and Doppler frequency for target and interference cells	EVA, 3 kmph, Use different channel seed for between cells	
CRS configuration	2 CRS ports with planning (non-colliding)	
CSI-RS configuration	None	4 CSI-RS ports, and 5 ms periodicity
PMI for target signal	Wideband PMI	
H-ARQ	8 HARQ processes and max 4 transmissions	
PCFICH	CFI = 2	
Resource allocation	50 RBs	

downlink transmit modes is listed in Table 3.1 in Chapter 3 and the simulations assumptions are outlined in Table 12.1.

Figure 12.5 shows the Throughput (in Mbps) versus SINR of MMSE-MRC and MMSE-IRC receiver for MCS levels 7, 8 and 9 defined in [3] for transmit mode-6 using two transmit

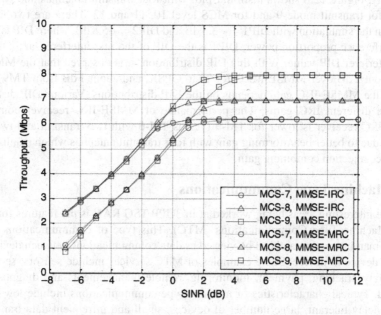

Figure 12.5 Throughput (Mbps) versus SINR for TM6 MCS-7,8,9 for MMSE-MRC and MMSE-IRC receiver using two transmit and two receive antennas.

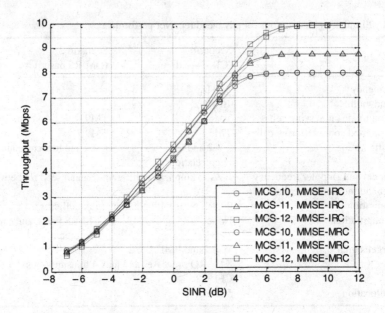

Figure 12.6 Throughput (Mbps) versus SINR for TM9 MCS-10,11,12 for MMSE-MRC and MMSE-IRC receiver using four transmit and two receive antennas.

and two receive antennas at eNodeB and UE respectively. It is assumed that the correlations of the signal and the interference + noise are both low.

Similarly, Figure 12.6 shows a similar plot with four transmit antennas and two receive antennas for transmit mode-9 and for MCS level 10, 11 and 12. There are two interferers assumed in the simulation with DIP1 = −3.3 dB and DIP2 = −5.8 dB, where DIP is the dominant interference proportion power, DIP1 is the DIP of the first interferer, and DIP2 is the second interferer DIP value. With this DIP distribution, it is observed that the MMSE-IRC gain is about 2 dB over MMSE-MRC for TM6 QPSK, and about 1 dB from TM9 16QAM. Note that the MMSE-IRC gain depends on the DIP distributions. Various DIP distributions shall have different IRC gains. The performance of MMSE-IRC receiver compared to MMSE-MRC receiver is lower for TM-9 (versus TM-6 with two transmit and two receive antennas) due to better beamforming gain with four transmit antennas which results in lower interference rejection combining gain.

12.4 Machine Type Communications

One of the interesting areas being worked at in 3GPP TSG RAN is the features for the support of Machine Type Communications (MTC). This type of communications does not require human interaction and can be viewed as data communications between devices and a server, or device to device. Some examples of MTC services include sensors, security and public safety, tracking, payment, monitoring, remote managements and diagnostics, and smart grid. Typical characteristics of machine type communications include low cost, low mobility, delay tolerant, large number of devices, small and infrequent data transmission, high reliability, time-controlled operation and group-based communications.

Table 12.2 Link budget comparison (uplink)

Uplink	LTE (Rel-8)	UMTS (HSUPA)	CDMA1x (1 × EV-DO)	GSM
UE EIRP				
Power (dBm)	23	23	23	33
Tx Antenna Gain (dBi)	−2.0	−2.0	−2.0	−2.0
EIRP (dBm)	21.0	21.0	21.0	31.0
eNB Receiver Sensitivity				
Receiver Antenna Gain (dBi)	17.0	17.0	17.0	17.0
Cable + Connector Loss (dB)	1.0	1.0	1.0	1.0
Noise Figure (dB)	5.0	5.0	5.0	5.0
Thermal Noise (kT) (dBm/Hz)	−174.0	−174.0	−174.0	−174.0
Bandwidth (KHz)	360.0	3840.0	1250.0	200.0
Required SINR (dB)	−7.1	−17.2	−12.4	7.0
Receiver Sensitivity	−136.5	−136.3	−136.4	−125.0
Margins				
Soft Handoff Gain (dB)	0.0	2.0	2.0	0.0
Lognormal Fade Margin (dB)	5.6	5.6	5.6	5.6
Interference Margin (dB)	1.0	3.0	3.0	1.0
Building Penetration Loss (dB)	18.0	18.0	18.0	18.0
Body Loss (dB)	2.0	2.0	2.0	2.0
Total System Margin	26.6	26.6	26.6	26.6
Maximum Allowable Path Loss	130.9	130.7	130.8	129.4

With the widespread introduction of LTE and decommissioning of legacy systems, migration of MTC devices to LTE is under investigation by cellular operators. To ensure that there are no coverage holes with such migration, LTE must provide the same or larger system coverage compared to legacy systems such as GSM or UMTS. This will allow LTE deployment using existing cellular sites. Typically, system coverage is limited by the uplink due to the limited transmission power of the device. Table 12.2 provides uplink link budget comparison of LTE to several cellular systems (UMTS, CDMA1x, and GSM) that are currently used for machine type communications [4]. From the Table 12.2, it is seen that LTE provides comparable link budget to existing cellular systems. Thus, the coverage hole is not expected as legacy cellular systems are decommissioned and replaced with LTE.

In general, MTC services are uplink-centric as information is mostly transmitted from the MTC devices to the network. Downlink transmissions may be in the form of control messaging, acknowledgements, and updates which may be done in a group-wise manner (e.g., using group-addressing or MBSFN concepts). From [5], it is seen that, in the uplink, Release 8 LTE provides a capacity gain of approximately 2.5 times over HSPA, and six times over GSM. As a result, MTC capacity, defined as the number of supported devices within an area, will be significantly enhanced with LTE.

Figure 12.7 illustrates the uplink capacity for low data-rate health-care monitoring sensor devices. The traffic model is that of a home monitoring service where the device transmits

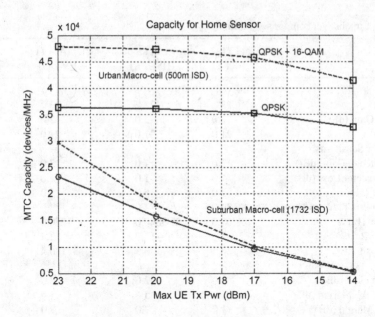

Figure 12.7 Uplink capacity for low data-rate home sensor devices [4].

128 bytes of information every 1 min (equivalent to a data rate of 17 bits/s). Two scenarios, urban and suburban macro-cell deployment, are evaluated. For urban macro-cell scenario, the eNodeB inter-site distance is 500 m, while for suburban macro-cell it is 1732 m. From the figure, it is seen that a substantial number of devices can be accommodated. When normalized by the system bandwidth, approximately 30 000 devices per MHz can be supported in suburban macro-cell and 48 000 devices per MHz can be supported in an urban macro-cell.

Some simple methods have been added to protect network overload in some specific cases and, on the other hand, a 3GPP study is ongoing to look at how to make the machine-to-machine modem more cost efficient compared to the to Release 8 Category 1 UE. This work is motivated especially by the low cost level of GSM based solutions for machine-to-machine communications. The items to be looked at include:

- Support of more than one receiver antenna, as the Release 8 LTE baseline requires UEs to support two-antenna reception as the minimum UE capability thus mandating the use of two RF chains and antennas. By eliminating one receive RF chain, significant RF cost saving can be achieved. In addition, this also simplifies the baseband processing requirements since spatial or diversity processing is not supported. As a result, component costs can be reduced due to simpler processing in component blocks such as channel estimation and smaller memory requirement. From a performance perspective, although there is a significant loss of capacity on the downlink when only one RF receiver chain is present, the downlink spectral efficiency is still greater than the uplink spectral efficiency. Thus, there should be no impact to MTC capacity as it is typically uplink limited. However, although there is no impact to MTC capacity there is an impact to system capacity for mixed usage as MTC devices take more downlink resources. The loss depends on how much of the downlink system resource MTC devices takes up and how heavily loaded the system is. From a coverage perspective, it has been shown that LTE coverage is still limited by the uplink even if only one receive antenna is available. Thus, there should be no impact to coverage.

- Support of 20 MHz bandwidth, to see what would be the UE cost savings with reduced bandwidth compared to the necessary modifications on the network side to accommodate such a UE in a network using larger bandwidth than the UE capability supports. The majority of the saving is expected to be from the baseband module where lower-cost processor, FFT, ADC/DAC, UL/DL channel processing units, and memory can be used. In the uplink it is possible to provision the bandwidth such that MTC devices can be confined to a portion that is smaller than the system bandwidth. In the downlink, however, it is not possible to provision a narrow bandwidth for MTC since the common control channels (PHICH, PCFICH, and PDCCH) span the entire downlink bandwidth. As a result, specification changes will be required to enable a narrow-band low-cost MTC UE to operate on a wideband carrier.

- Reduction of the peak data rate. The currently supported smallest data rates (with UE category 1 as covered in Chapter 3) are 10 Mbps for downlink and 5 Mbps for the uplink directions. In some of the MTC solutions such as metering applications the amount of information to be transmitted is very limited and may not benefit from such a high data rates in the radio modem. Several methods may be used to reduce both downlink and uplink peak rates, including QPSK-only modulation, reduced TBS subset, and reduced number of HARQ processes. This can result in some saving in the baseband and RF modules. Some methods (e.g., reduced TBS subset and reduced number of HARQ processes) have no impact on system capacity and coverage, and require only minor modifications to eNodeB implementation. Other methods such as QPSK-only modulation can affect system performance. From Figure 12.7, it can be seen that the capacity loss with QPSK-only modulation in the uplink is 19% for suburban macro-cell scenario and 24% for urban macro-cell scenario.

- The UE transmit power reduction is also seen as one possibility for smaller cost, with the current power level required in the devices being 23 dBm. The reduction of the power will have potential impacts to uplink range which could be considered to be compensated by mean such HARQ and sub-frame bundling. From Figure 12.7, it can be seen that the capacity loss is small for urban macro-cell scenario where the cell size is small and the system is interference limited. In this case, a 6 dB reduction in the maximum output power results in only a 5% decrease in uplink capacity. On the other hand, for suburban macro-cell deployment where the cell size is large and the system is noise limited, capacity loss is significant. A 6 dB reduction in the maximum output power results in 65% decrease in uplink capacity. Likewise, coverage holes may be introduced for large cells in existing networks. The loss is proportional to the amount of power reduction.

- Half duplex operation is also often mentioned as one of the aspects contributing to the low cost of GSM based modems. If one would operate FDD LTE in such a mode that uplink and downlink would not need to be transmitting/receiving simultaneously (like in TDD) the resulting half duplex UE would not need duplex filters for the bands in questions. From the network side this especially requires the eNodeB scheduler to accommodate the resulting limitations in the UE transmission and reception capabilities. In this case, the scheduler must ensure there are no scheduling conflicts for half-duplex UEs. This will require the scheduler to consider data and control traffic in both directions when making scheduling decisions. For example, the downlink scheduler must know of current uplink transmission (e.g., CQI, ACK/NACK, semi-persistent scheduled PUSCH, or PUSCH retransmissions). Likewise, the uplink scheduler must be aware of upcoming downlink ACK/NACK or semi-persistently scheduled transmission. This can add to the scheduler complexity significantly. Furthermore, like in TDD, a switching time will be required by half-duplex UE when transitioning from one direction to another. This may require changes to the specifications (e.g., to add guard periods or symbols), or through further scheduling restrictions to ensure that there is sufficient gap (e.g., no consecutive transmission when switching).

Table 12.3 lists approximately saving in the UE modem from various techniques described previously. From the table, it is seen that single receive antenna, half-duplex FDD, and reduced bandwidth provide the most saving. Having only one receiver RF chain can provide an estimated 14–18% cost saving for a reference LTE modem. There is a slight impact on system capacity and coverage, but a meaningful reduction in power consumption. Low-cost MTC UE should therefore be equipped with only a single receive RF chain. Table 12.3 shows the estimated modem cost saving relative to Release 8 Category 1 UE, with the Release 8 UE categories introduced in Chapter 3.

Although half-duplex operations can provide some saving, they require significant implementation changes in the network infrastructure. Therefore, the half-duplex operation should not be a mandatory requirement for low-cost MTC devices. For a reduction in maximum bandwidth, specification changes have to be introduced and their impacts should be carefully considered before this technique is adopted.

Reducing the maximum output power can result in 1–3% cost saving. In large cells, however, capacity loss can be significant and coverage holes may be introduced. As a result, the marginal cost saving does not necessary justify this reduction. Peak rate reduction can result in 5–7% cost saving. Some methods have no impact on system capacity and coverage, and require only minor modifications to eNodeB implementation. These methods could thus be adopted for low-cost MTC devices, with the specification changes to be concluded within the expected 3GPP MTC type UE work item expected to start in 2H/2012 following the study phase.

In Reference [6] it has been also estimated that roughly 40% of the total modem cost come from RF parts and 60% from the baseband. Within the RF parts the RF transceiver dominates

Table 12.3 Saving in UE modem for various techniques [4]

Technique	Approximate Saving (%)
Single receive antenna	14–18%
Reduced maximum UE power	1–3%
Half-Duplex FDD	9–12%
Reduced maximum UE bandwidth	6–10%
Peak rate reduction	5–7%

Table 12.4 Division of costs within RF parts [6]

Component	Part of total RF cost (%)
RF amplifier	25–30%
RF transceiver	40–50%
Duplexer/switch	15–25%
Other	0–10%

the costs (with the cost including the mixers, local oscillator and Low-Noise Amplifier (LNA) as shown in Table 12.4. From the evaluations for the distribution of the baseband costs there is a lot of variations in [6] depending on the source, but the largest is the receiver processing block which runs in some estimates up to 40% of the total baseband costs when including the MIMO related processing, clearly ahead of the Analog-to Digital Conversion (ADC)/Digital-to-Analog Conversion (DAC), FFT/IFFT, Turbo decoding, HARQ buffering and other baseband functionalities.

12.5 Carrier Aggregation Enhancements

Following the introduction of the carrier aggregation in Release 10, the work item [7] was set up to address improvements for the carrier aggregation operation. As part of the work item, the following new items were worked to be part of Release 11, explained in more detail later in this section:

- Support of multiple uplink timing advance values;
- Intra-band non-contiguous carrier aggregation;
- Support of inter-band carrier aggregation for TDD DL and UL including different uplink-downlink configurations on different bands;
- New carrier type (this to be completed in Release 12 only).

The support of multiple uplink timing advance values is motivated to enable operation in such a scenario where band specific repeaters are in use, as discussed in Chapter 5. If the repeater is available only on one of the bands, the resulting uplink propagation time is different between different frequency bands due to the internal delay of the repeater and thus a single timing advance value is not sufficient with the use of two uplinks. The MAC layer signaling in Release 11 will allow the eNodeB to provide the UE timing advance values specific for each uplink band..

As part of the Release 11 work is also the framework for intra-band non-contiguous carrier aggregation operation. With the intra-band non-contiguous operation there can be another operator or system operating in between the carriers thus the devices must have enough protection to cover such case with up to 33 dB adjacent carrier attenuation as defined in [8] and illustrated in Figure 12.8. As the duplex filter only filters the full band, then the UE needs additional filters or rather two full receiver chains as shown in Figure 12.9. The figure assumes single transmit antenna (no uplink MIMO) and use of single uplink carrier only.

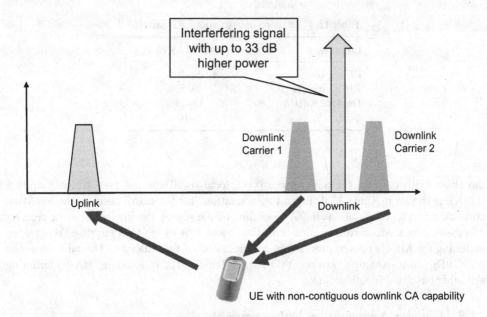

Figure 12.8 A downlink non-contiguous intra-band carrier aggregation interference scenario.

Support of inter-band carrier aggregation for TDD DL and UL includes different uplink-downlink configurations on different bands. This means that TDD UE supporting such a carrier aggregation case becomes a UE with full duplex capability to enable transmission and reception to happen simultaneously on different part of the frequency spectrum, like is done on the FDD side.

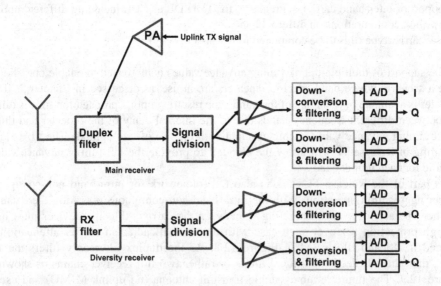

Figure 12.9 Example of a received structure for the non-contiguous intra-band CA operation.

The most significant new layer 1 feature related to the carrier aggregation operation in Release 11 work is the new carrier type. The work on new carrier type is focused on the downlink and no major changes to the uplink operation are expected. New, non-backwards compatible carrier types were already considered during the Release 10 standardization, but at that time they were decided to be not included in the specifications. The main candidates identified in Release 10 were:

- **The Extension Carrier:** A carrier having no common reference signals, PSS/SSS, PDCCH/PHICH/PCFICH or PBCH present. Mobility is based on other, backwards compatible carriers, that is, the extension carrier would be an SCell.
- **Carrier Segments:** A contiguous bandwidth extension of a backwards compatible carrier.

In Release 11 the discussion on new carrier type continued and an agreement to introduce new carrier type was reached. The Release 11 new carrier is expected to operate as a SCell so that, for example, mobility procedures can rely on the backwards compatible PCell. Furthermore, the new carrier type should support synchronized and non-synchronized carrier aggregation scenarios, that is, both inter- and intra-band carrier aggregation.

The main motivations for the new carrier type are:

- **Enhanced spectral efficiency:** This can be achieved primarily through reduced signaling as well as reference signal overhead.
- **Improved support for HetNet:** The possibility of not transmitting many of the common channel or signals reduces interference and helps in tackling many of the issues the work on HetNets is dealing with.
- **eNodeB energy efficiency:** Having a possibility to dynamically mute the at least some of the sub-frames completely when there is no data traffic allows for switching off the eNodeB transmitted to save power.

In the absence of many of the Release 8 common signals, the system operation with new carrier type relies heavily on the Release 10 MIMO functionality, namely Transmission Mode 9 with DM RS used for data demodulation and CSI-RS for channel state information feedback. Another Release 11 feature, ePDCCH, plays also a key role in the new carrier type as it together with cross-carrier scheduling allows for removing the common downlink control channels (PDCCH, PHICH, PFCICH) from the new carrier type.

The main items for the new carrier type have been resolved, including maintaining reduced CRS content (periodically every 5 ms) while otherwise using the CSI-RS and DM-'RS for feedback generation and demodulation purposes, as illustrated in Figure 12.10. New carrier type will be completed in Release 12.

12.6   Enhanced Downlink Control Channel

The work on the enhanced downlink control channel is referring to the enhanced PDCCH (ePDCCH) to improve the capabilities and performance of the physical layer control signaling, with the actual work being started from the beginning of 2012, following earlier studies. The work focuses on the downlink and the related changes to the uplink operation are expected to be minor. The key targets for the work are to enable:

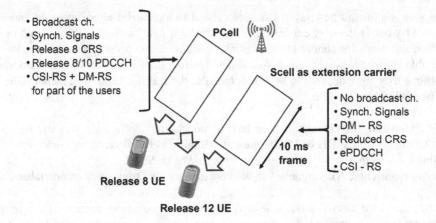

Figure 12.10   New carrier type in Release 12.

- Support for higher control channel capacity, which has linkage to the scenarios of the MTC with very large number of low data rate UEs accessing the network.
- Support for frequency domain scheduling and ICIC for downlink control channel.
- Support for beamforming. Due to transmit diversity operation, in Release 10 the downlink control channels are the only channels not benefitting from the increased number of eNodeB antennas in terms of beamforming gain (downlink) or RX diversity (uplink).

As introduced in Chapter 3, the Release 8 based PDCCH is scheduled over the full carrier bandwidth which prevents from performing frequency domain scheduling and interference avoidance for the physical layer downlink control signaling in a similar way as can be done for the actual user data. Furthermore, with more than two eNodeB transmit antennas the PDCCH is transmitted using transmit diversity, which while providing robust system operation on average, is clearly suboptimal especially when the number of eNodeB transmit antennas is great.

The main motivation in the ePDCCH work in Release 11 is to allow the control signaling to benefit from the same gain mechanisms as the data: frequency domain scheduling and beamforming. Assuming that somewhat accurate channel state information is available to enable ideal scheduling and link adaptation decisions for data in terms of frequency (PRB) as well as spatial (precoding) domain, obviously the same scheduling decision should be the optimal for the control signals too.

The ePDCCH in Release 11 is planned to meet these targets by enabling similar type of frequency domain scheduling as for the data where a UE is getting control signaling localized within particular physical resource block(s) as illustrated in Figure 12.11. This allows for utilization of the CSI feedback for the benefit of the control information and also opens new possibilities for interference avoidance as now control information transmission can be restricted to take place over only a limited number of the physical resource blocks. As visible in Figure 12.11, the existing control channels (PDCCH) may remain unchanged to serve the UEs based on Release 10 or earlier. Similarly, pre-coding can be applied on the control signals as well.

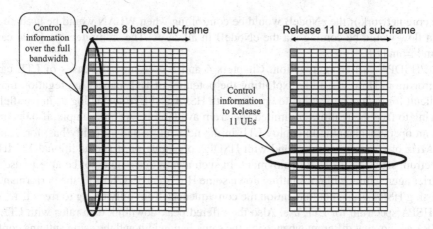

Figure 12.11 Introduction of the frequency domain scheduling capability for downlink control information.

Besides the targets listed above, the necessary elements related to other Release 11 enhancements are to be covered as well, such as support for the new carrier type in terms of new signaling needs as well as accommodating the additions due to the introduction of CoMP operation.

12.7 Release 12 LTE-Advanced Outlook

3GPP will determine, from mid-2012 onwards, the overall framework for the Release 12 work, starting from the 3GPP TSG RAN workshop scheduled for mid-2012 addressing work from 3GPP TSG RAN Release 12 and beyond, with some of the topics addressed already earlier in 3GPP presented here. It is worth noting that the Release 12 work program is not yet concluded and is expected to contain several additional topics to the expected items listed in the following. The topics postponed from Release 11 include:

- Device-to-device communication is a topic with work on-going in 3GPP already in the requirements domain from 2H/2011 onwards. The different use cases include close proximity communications for such users that would have large amount of data to transmit to users close by. Another type of use case is the use of LTE for public safety where direct communication between terminals would also be desired in addition to the regular cellular type of communications. The RAN level study item proposal can be found from [9].
- LTE and WLAN dynamic path switching (referred also as LTE/WLAN carrier aggregation) looks for tighter interworking than the existing core network off-loading solutions. The key motivation factor with integration in the radio level of WLAN and LTE is performance. The faster the WLAN use can be control, the more reliable and with better performance (of combined system performance) the joint operation can be carried out. 3GPP has defined in earlier releases different core network based solutions for the WLAN use, but none of those has been so far widely deployed in the market place. With the radio level integration it would be possible to use LTE and WLAN together without impacting

the core network if the eNodeB would be controlling when WLAN would be used and the data flow would be coming via the eNodeB in all cases. The study item proposal can be found from [10].

- LTE/HSDPA carrier aggregation. Chapters 4 and 5 presented the cases of LTE carrier aggregation. However, fully exploiting the potential of LTE carrier aggregation may be difficult for operators needing to operate both HSPA and LTE technologies in parallel and having to face the realities of limited spectrum availability. As an example, if one considers an operator planning to deploy LTE in the near future and currently has, for example 15 MHz of spectrum on one band with HSDPA operating on it and additional 15 MHz of spectrum set aside for LTE deployment. In such a case one would not be able to use LTE carrier aggregation without shutting down some HSPA carriers, hurting the performance of existing HSPA users, not to mention the consequences from fully trying to free all 15 MHz of HSPA spectrum for LTE use. Also the offered peak downlink data rates with LTE and HSPA are not that different when using the same bandwidth and the same antenna configuration. The marketing of the new LTE network may pose a challenge if the end user peak data rates remain practically unchanged. The approach with the LTE/HSDPA carrier aggregation has the following performance benefits for an operator as presented in [11]:
  - Allows one to combine the peak data rates of both radio systems reaching this data rates in the order of 100 Mbps with 10 MHz of LTE and 10 MHz of HSDPA spectrum.
  - Provides dynamic load balancing between the two radios as similar to the LTE carrier aggregation one case share on the TTI level the data between HSDPA and LTE downlink carriers.
  - Ensures highest possible spectrum utilization when both LTE and HSPA systems are deployed with the joint scheduling approach as in the LTE only carrier aggregation.
  - Allows to reach the carrier aggregation benefits in addition to the peak data rate also for the cell edge data rate.
- The LTE/HSDPA carrier aggregation shown in Figure 12.12 assumes the use of single uplink to avoid impacts to the achievable uplink range and to limit the resulting implementation complexity. This is in-line with the practical Release 10 based LTE carrier aggregation work, where first the focus is in the case of downlink only carrier aggregation and the RF requirements will be done only later to cover the two uplink cases. The study item proposal on LTE/HSDPA carrier aggregation can be found in [12].

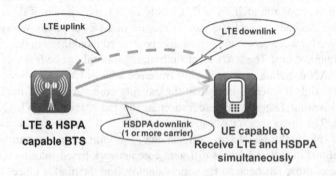

Figure 12.12 An LTE/HSDPA carrier aggregation operation with single uplink.

From Release 11 studies on downlink MIMO enhancements a further area identified was to continue work on the Channel State Information (CSI) feedback enhancements. The Release 12 work (assumed to start September 2012 onwards) will focus on looking at the improved feedback and potentially new codebooks which would provider finer spatial and/or frequency domain granularity improving the capacity especially for the 4TX MIMO case [13]. Further work areas are covered in [14].

The 3GPP Release 12 schedule is not yet officially finalized but with the work program to be decided during the second half of 2012, the expected finalization is to take place in the second half of 2014 as mentioned in Chapter 2. The Release 11 content is to be in place by September 2012 to enable the protocol freezing at the end of 2012 or early 2013.

12.8 Conclusions

In this chapter, we have covered the LTE-Advanced Release 11 and 12 features and outlook. The Release 11 work will in many respects complement the Release 10 LTE-Advanced framework while Release 12 is foreseen to contain more completely new items with some of them introduced in this chapter. The Release 11 topic list is more or less concluded with the items to be ready by the end of 2012, then followed by the Release 12 work and study items expected to be finalized during 2014 with the official timing still to be determined in 3GPP. Many of the Release 11 topics enable further improved performance of the LTE and LTE-Advanced operation, as well as improved deployment flexibility in different network topologies foreseen to appear to meet the increased demand for mobile data capacity and data rates.

References

1. 3GPP Tdoc R4-113528 (June 2011) Performance of Interference Rejection Combining Receiver for LTE NTT DOCOMO.
2. 3GPP Tdoc R4-121029 (February 2012) TP for Enhanced performance requirements for LTE UE SI, NTT DOCOMO.
3. 3GPP TS 36.213 (January 2012) Physical layer procedures, Table 7.1.7.1-1, v10.4.0.
4. Ratasuk, R.,Tan, J. and Ghosh, A.(2012)Coverage and capacity analysis for machine type communications in LTE.IEEE Vehicular Technology Conference, May 2012.
5. Holma, H. and Toskala, A.(2011) *LTE for UMTS*,2nd edn, John Wiley & Sons, Ltd.,Chichester.
6. 3GPP TR 36.888 (February 2012), Study on provision of low-cost MTC UEs based on LTE, Version 1.0.0.
7. 3GPP Tdoc RP-111115 (September 2011), LTE Carrier Aggregation Enhancements WID, Nokia Corporation.
8. 3GPP Technical Specification TS 36.101 (March 2012) UE transmission and reception.
9. 3GPP Tdoc RP-111093 (September 2011) Study on LTE device to device discover and communications – radio aspects, Qualcomm.
10. 3GPP Tdoc RP-111354 (September 2011) New study item proposal for dynamic flow switching between 3GPP-LTE and WLAN, 1 Intel.
11. 3GPP Tdoc R1-111060 (February 2011) Aggregating HSDPA and LTE carriers, Nokia Siemens Networks, Nokia.
12. 3GPP Tdoc RP-120120 (February 2012) New study item proposal for LTE and HSDPA Carrier Aggregation, Nokia Siemens Networks, Nokia.
13. 3GPP Tdoc RP-120413 (February 2012) Further Downlink MIMO Enhancement for LTE-Advanced, Alcatel-Lucent, Alcatel-Lucent Shanghai Bell.
14. 3GPP Tdoc RWS-120045 (June 2012) Summary of TSG-RAN workshop on Release 12 and onward, TSG-RAN Chairman.

13

Coordinated Multipoint Transmission and Reception

Harri Holma, Kari Hooli, Pasi Kinnunen, Troels Kolding, Patrick Marsch and Xiaoyi Wang

13.1 Introduction

The peak data rates provided by LTE radio are substantially higher than the average data rates or the cell edge data rates – the difference can be a factor of 10 or more. The main reason is inter-cell interference, which limits signal-to-interference ratios and consequently also achievable data rates. While GSM systems apply frequency reuse (for example reusing the same resources only every seven or 12 cells), LTE uses a frequency reuse factor of one, meaning that inter-cell interference is particularly high. While this interference has traditionally been minimized by careful RF planning with antenna selections, antenna tilt and parameter settings, LTE radio brings further tools to mitigate interference in frequency and time domain in 3GPP Releases 8 and 10, as discussed in Chapter 8. A completely different view on multi-cell interference paths is now being enabled through the option of coordinated or cooperative signal processing across multiple cells. Such techniques, often referred to as Coordinated Multipoint (CoMP) Transmission and Reception and addressed in a work item for LTE Release 11 in 3GPP, are able to exploit interference paths between cells, rather than seeing them as problematic. Consequently, they can lead to higher capacity per area and, more importantly, offer an increased and homogeneous quality of service across cell areas. CoMP concepts, network architecture implications, and performance gains are described in this chapter.

13.2 CoMP Concept

CoMP is based on coordinated or cooperative signal processing among multiple cells. Let us start with the most sophisticated CoMP concept, namely downlink multi-cell Joint

LTE-Advanced: 3GPP Solution for IMT-Advanced, First Edition. Edited by Harri Holma and Antti Toskala.
© 2012 John Wiley & Sons, Ltd. Published 2012 by John Wiley & Sons, Ltd.

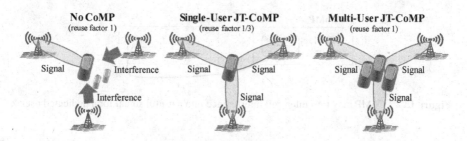

Figure 13.1 Joint downlink transmission from multiple cells.

Transmission (JT), as an illustrative example of how CoMP changes the view on interference in mobile communications.

On the left side of Figure 13.1, we see three cells simultaneously transmitting in a conventional (i.e. non-coordinated) way to three terminals on the same physical resources. For each of the terminals, this means that the desired signal from the assigned cell is impaired by interfering signals from the other two cells. In JT CoMP, one could now let all three cells transmit jointly to one of the terminals on the same resources in time and frequency. This may sound a bit like soft handover from Wideband Code Division Multiple Access (WCDMA), where two cells send the same signal with different scrambling codes to one terminal. In both CoMP and soft handover, the result is that signals which formerly were inter-cell interference are now useful signals adding up in at the receiver side. As shown in the centre plot of Figure 13.1, when certain physical resources are now transmitted towards the same user from different cells in single-user mode, the system becomes effectively a dynamic reuse factor 1/3 system as a single CoMP user now blocks resources in three cells. This is a similar effect as exploited with Inter-Cell Interference Coordination (ICIC) mechanisms, but instead of muting resources, these are used constructively to boost a certain user's Signal-to-Noise-Ratio (SINR). In high load macrocells, the loss of lowering the effective reuse factor may sometimes offset CoMP gains and to mitigate this issue, sets of users may be formed as shown on the right hand side, allowing for multi-user CoMP transmission where users share the same physical resources. For this case, the total transmission power per user per physical resource block still remains the same as in the no CoMP case, while inter-cell interference may be effectively overcome.

The benefit of such a joint downlink transmission can be illustrated in a simplified way by using the so-called Shannon formula where the channel capacity (C) is a logarithmic function of SINR. In the stated case of JT CoMP, all interference is (in an idealistic case) turned into useful signal energy, and hence appears in the nominator of the SINR term instead of the denominator, as shown in Figure 13.2, leading to a higher capacity [1].

The commonality between soft handover and JT CoMP is that the base stations redundantly send the same data to the terminal, aiming at improving the SINR. Both solutions target for increased throughput and more robust handover by multi-cell transmission. In addition, JT CoMP exploits channel feedback from the terminal side in order to achieve a coherent overlap of signals at the receiver side, yielding a so-called beamforming gain. This gives an additional SINR gain, and consequently a further increase in data rates.

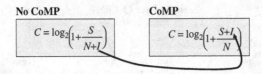

Figure 13.2 CoMP may turn inter-cell interference into a useful signal for a dedicated user.

In addition to JT schemes, downlink CoMP also comprises schemes where cells simply coordinate the transmissions to different users in the spatial domain, or where beamforming is performed such that inter-cell interference is mitigated. The complete range of downlink CoMP schemes will be introduced in Section 13.4 and that of uplink CoMP schemes in Section 13.5.

Let us now state what differentiates CoMP from other interference coordination techniques in LTE, as summarized in Table 13.1. Release 8 allows avoiding inter-cell interference in frequency domain by minimizing the usage of those resource blocks posing strong interference. Release 8 Inter-cell Interference Coordination (ICIC) can be obtained by frequency dependent reporting from the User Equipment (UE) or by signalling between base stations over the X2 interface. Release 8 ICIC does not require any centralized control nor base station time synchronization. Release 10 brings Enhanced ICIC that avoids inter-cell interference in the time domain by using synchronized base stations. Enhanced ICIC is designed especially for heterogeneous networks to enable co-channel deployment of macro-cells and small cells, as discussed in Chapter 8. CoMP can be considered as the next step in the management of inter-cell interference. Intra-site CoMP can be done within one eNodeB without any requirements for the transport network, while inter-site CoMP sets high requirements for the transport network to deliver the samples between RF head and a centralized baseband location.

Table 13.1 Interference coordination options

	ICIC	Enhanced ICIC	CoMP
Operating domain	Frequency domain	Time domain	Spatial domain
Base station time domain synchronization	Not needed	Yes	Yes
Transport requirements	Low requirements since only control plane carried	Low requirements since only control plane carried	High requirements in Joint Transmission/Reception case since also received samples are carried
Time scale of optimization	Relatively slow (>10 ms)	Relatively slow (>10 ms)	Fast (1 ms) in Joint Transmission/Reception case
3GPP Release	Release 8	Release 10	Release 11

13.3 Radio Network Architecture Options

Release 8 LTE radio networks are designed to be distributed where all the radio network functionalities are located in the base station. Frequency domain inter-cell coordination in Release 8 can be obtained with a distributed architecture with slow information exchange between base stations, or even completely distributed algorithms based on the feedback from the terminals. More advanced real-time inter-cell interference coordination such as CoMP would benefit from a common and centralized control unit that could optimize the transmissions from multiple cells on a small time scale. Therefore, CoMP features can also have a major impact on the radio network architecture and the transport requirements. Figure 13.3 illustrates the two extreme architecture options: Distributed baseband and centralized baseband. The distributed baseband is the normal solution in LTE networks. The centralized baseband uses simple RF heads at the base station site connected over high speed fibre to the baseband pool, also called a baseband hotel.

Let's look quickly back at the history of radio network architecture. The original idea in WCDMA radio was to have the radio intelligence in the Radio Network Controller (RNC) while the base station only included RF and Layer 1 encoding and decoding. The transport requirements were relatively low due to low user data rates. High Speed Packet Access (HSPA) specifications pushed more functionality to the base station to allow faster scheduling and retransmissions. LTE specifications went further by relocating all the radio functionalities to the base station. We have seen gradual evolution from hierarchical and centralized architecture towards flat and decentralized. The baseband hotel solution is going to another direction where the RF head is simple and 'stupid' and all the control is located in the baseband hotel including Layer 1. The baseband hotel drastically increases the requirements for the transport because it is not only user symbols but radio samples that need to be transmitted from RF head to baseband hotel. The transport must support data rates beyond 10 Gbps and a total latency below one half of a millisecond, which implies that baseband hotels require dedicated fibre connection or high bandwidth microwave radio to the RF heads. That transport solution with baseband hotel is called fronthaul in order to differentiate it from the traditional transport which is called

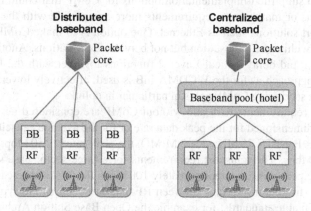

Figure 13.3 Distributed and centralized baseband (BB = Baseband, RF = Radio Frequency).

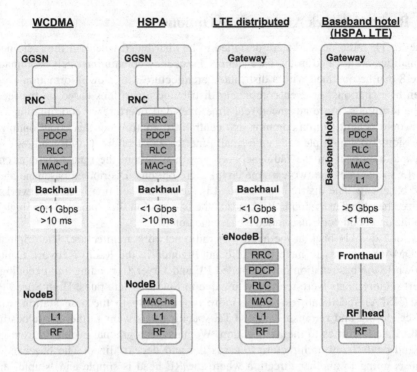

Figure 13.4 Radio network architecture options.

backhaul. The transport is behind (after) decoding in the traditional solution while the transport is in front of decoding in the baseband hotel. An overview of example architecture options is illustrated in Figure 13.4. 'Centralized controller' means an entity responsible for the transmissions across different cells and may be implemented both as a master-slave relationship between distributed but fast X2-interconnected baseband modules or as a centralized baseband architecture as shown in Figure 13.4.

There are also simplified implementation options for CoMP that could reduce the transport requirements or make those requirements more compatible with the performance of standard transport solutions such as Ethernet. One option is to make CoMP operation only between sectors within one base station but not between base stations. Another possibility is to locate Layer 1 and delay-critical Layer 2 functions together with the RF head. In this setup, a similar approach as for the WCDMA IuB is used, effectively lowering the transport requirements but sacrificing CoMP gains in particular in uplink.

The transport requirements with and without CoMP are considered here. The transport network can be dimensioned for the peak data rate or for the average capacity with a distributed architecture. For LTE 20 MHz 2 × 2 MIMO, the peak rate is 150 Mbps and typical cell capacity 30–40 Mbps. The transport requirements for a three-sector base station would be 450 Mbps for the peak case and approximately 100 Mbps for the average case. When CoMP is used, then the fronthaul interface between RF and baseband would typically be implemented using available standards; for example the Open Base Station Architecture Initiative (OBSAI) or the Common Public Radio Interface (CPRI) specifications. Principle is to transfer Imaginary Quadratic (IQ) samples, so the effective data rate depends on the system

Table 13.2 Example downlink IQ sample rates for fronthaul depending on RF site configuration. Actual configured rates depend on the actual segmentation of cells/sectors to RF and baseband modules

RAT	Bandwidth	Number of antennas	Number of sectors	Fronthaul rate (raw IQ samples)[a]
WCDMA	10 MHz	2	3	1.5 Gbps
LTE FDD	20 MHz	2	3	5.5 Gbps
LTE FDD	20 MHz	4	6	22.1 Gbps
LTE TDD	20 MHz	8	3	22.1 Gbps

[a] For for example, CPRI: overhead of ~7% is considered and nearest matching line rate used, for example, Nx0.6144 Gbps.

configuration, namely the core radio parameters, the system bandwidth, the amount of antennas and so on. This also means that requirements for transport increase with the peak rate evolution of the radio sites. Each interface specification has a set of core rates including control overhead. Example IQ sample rate requirements for different RF site configurations in the downlink direction are illustrated in Table 13.2. Similar requirements exist in the uplink direction but sometimes differ a bit, for example in the case of WCDMA.

It is obvious that costs of fronthaul have to be contained for the operator, either through the use of specific low-cost fibre technologies such as passive Coarse Wavelength Division Multiplexing (CWDM) on dark fibre or by confining the fronthaul to a shorter range. The required fronthaul data rate increases with larger bandwidth and with higher number of antennas. The fronthaul requirements can be 50 times higher than the typical backhaul bandwidth requirements.

In spite of the high fronthaul requirements, centralized architectures are nevertheless being considered for other reasons than having a most effective implementation of CoMP. Other example benefits of centralized baseband hotels include:

- The radio network software or hardware upgrades are simpler when the upgrades can be done in a few centralized locations instead of a large number of base stations. That can lead to lower network operability expense.
- The handovers can be faster when there is no need for X2 or S1 signalling within a baseband hotel. The lower impact of ping-pong effects further allows tightening cell selection effectively improving data rate performance near the cell edge.
- The amount of mobility signalling to the core network is reduced when the signalling is aggregated by the base station hotel.
- The baseband dimensioning can be minimized when the baseband processing resources are pooled between many RF sites.

The centralized baseband hotel is also referred to as Centralized RAN or even 'Cloud-RAN' where the thinking is that the baseband processing could be located in the network cloud in the same way like the server functionalities for the generic processing. It is, however, important to notice the differences between cloud computing and

Table 13.3 Differences between cloud computing and centralized RAN applications

	Cloud computing	Centralized RAN (Baseband hotel)
Client (base station) data rate	Mbps range, low activity, bursty	Gbps range, constant stream
Latency and jitter	Tens of ms	<0.5 ms, jitter in ns range
Life-time of information being processed/stored	Long (content data)	Extremely short (data symbols and received samples)
Allowed recovery time	Seconds range (sometimes hours)	Millisecond range to avoid network outage
Number of clients connected to single centralized location	Thousands, even millions	Tens, maybe hundreds

Centralized/Cloud RAN. Cloud computing allows tens of milliseconds of delay and requires data rates of a few Mbps in bursts while the requirements for Centralized RAN are far higher. Also, the lifetime for the information in the centralized location is relatively long in cloud computing, even up to hours; while the information in a Centralized RAN outdates in a few ms. Cloud computing has typically thousands or even millions of computer clients accessing the information while a centralized RAN has connections to tens or hundreds RF heads only. The differences between cloud computing and centralized RAN are illustrated in Table 13.3.

13.4   Downlink CoMP Transmission

We now want to discuss the variants of downlink CoMP transmission, which differ substantially in the way multiple cells are involved in the transmission to single or multiple terminals. The different methods can be combined depending on how cells are clustered into CoMP sets. CoMP set refers to those cells from where a UE can receive or send data. A CoMP set can be considered similar to the active set in WCDMA. The two fundamentally different variants are illustrated in Figure 13.5. The centralized controller means an entity responsible for the transmissions across different cells and may be implemented both as a master-slave relationship between distributed but fast X2-interconnected baseband modules and as a centralized baseband architecture.

In *Coordinated Scheduling and Beamforming* (CS/CB), terminals receive a data transmission from their respective serving cell only, but adjacent cells perform coordinated scheduling and precoding such that inter-cell interference is avoided or mitigated to some extent. This can be considered as an extension of Release 8 inter-cell interference coordination (ICIC). In this way, the SINR is improved for the stated terminals and the scheduling freedom is reduced for the neighbour cells, how much depends on which antenna features exist in the system as well as the currently active user set.

In *Joint Processing* (JP), the network infrastructure provides the data to be transmitted to certain terminals to multiple cells, so that a larger extent of radio resource management flexibility is enabled allowing non-serving cells to participate in the transmission towards a certain UE. More specifically, this allows a fast and channel dependent switching of the

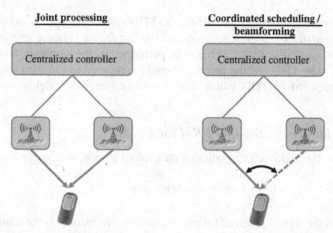

Figure 13.5   Joint processing and coordinated scheduling and beamforming.

(subset of) cells that transmit to a particular terminal on a particular radio resource. One here further distinguishes between:

- **Dynamic Cell Selection** *(DCS)*, where on each physical resource only one cell at a time may be transmitting to one terminal, and
- **Joint Transmission** *(JT)*, where multiple cells jointly and coherently transmit to one or multiple terminals on the same resource implementing the full advantages discussed in the beginning of this chapter.

In the following, various implementation aspects related to supporting downlink CoMP are described.

13.4.1   Enablers for Downlink CoMP in 3GPP

While for CS/CB, each involved cell only requires course channel state information (CSI) for the links from itself to the terminals in its own and adjacent cells, JT requires the compound channel among all involved cells and terminals within a so-called CoMP set to be known by all involved base stations. For this, the reference signal structure was optimized in Release 10. The common reference signals for Channel State Information (CSI) and user specific reference signals for decoding were separated. That optimization is useful for CoMP to provide the feedback information to the network and to allow user specific beamforming in the downlink. The details of the new reference signals are described in Chapter 6.

There are some differences between Time Division Duplex (TDD) and Frequency Division Duplex (FDD) from CoMP point of view. In the TDD case, uplink and downlink channels are reciprocal which means that the required downlink CSI can be obtained from uplink Sounding Reference Signals (SRS). In an FDD system, the uplink and downlink transmissions are separated by Duplex spacing, and the fast fading is hence uncorrelated. In this case, the estimation of downlink CSI has to be performed by the terminals, and then fed back to the network side. TDD has the potential benefit compared to FDD because no feedback signalling is required in TDD which allows avoiding feedback delay and quantization limitations.

13.4.2 Signal Processing and RRM for CoMP

For all schemes, the multi-cell transmission on a subset of system resources can be stated as

$$r = HWs + n, \tag{13.1}$$

where r denotes the signals received by one or multiple terminals, H is the multipath channel, W is a matrix containing the transmission antenna weights, s are the transmitted signals and n is noise. In the cases of CS/CB and DCS, W contains many zeros, due to the constraint that only one cell at a time may actively transmit to one user. In the JT CoMP case, all elements may be used, and there are two popular variants of designing the transmission antenna weights: One is based on the inversion of the multipath channel, which is known as a zero-forcing solution and is (ideally) able to completely remove the interference between the simultaneous transmissions to multiple terminals. The other variant is to use a so-called Wiener filter that finds the best trade-off between optimizing the desired signals as received by the terminals and minimizing interference. The antenna weight design for these two variants can be formally stated as

$$W = \beta \cdot H^H \left(HH^H \right)^{-1} \text{ or } W = \beta \cdot \left(H^H H + \frac{tr\{R_{nn}\}}{P} I \right)^{-1} H^H, \tag{13.2}$$

respectively, where β is a scaling factor that assures that an overall power constraint is met, R_{nn} is a diagonal matrix related to the noise or interference from outside the CoMP set as seen by the terminals, and P is the sum power to be invested into the joint transmission to the terminals.

A challenge particularly inherent in downlink CoMP is that adjacent cells have to schedule terminals onto resources in a coordinated way, so that CS/CB, DCS and JT can be applied effectively. Therefore, the neighbouring cells need to sacrifice scheduling flexibility to build so-called CoMP patterns. This means that there is a compromise in terms of how many cells are included in the CoMP set for a certain user, as the scheduling diversity then reduces to the point where the benefit of some users is paid by a too high price in terms of total system loss. A good CoMP implementation dynamically assesses these tradeoffs and bases its decisions on them. A fundamental limit to how many cells can be involved in CoMP transmission or reception is based on the network architecture limitations, for example how many cells are confined within the overall group of cells out of which CoMP sets can then be picked. We will refer to this group of cells as a CoMP *cluster*.

However, the key complexity of CoMP techniques in the downlink is that the CoMP sets should be determined optimally on a per-user basis, which means that a large number of cells should be considered as part of the CoMP cluster to get optimum benefits. CoMP gains

Table 13.4 Mechanisms for downlink CoMP gains

	User in downlink CoMP	Other users
Gain mechanisms	Additional transmit power by using power from multiple cells towards one UE	Users in CoMP set may be served faster reducing inter-site interference in low-load scenario
	Additional antenna diversity on top of existing diversity provided by MIMO and opportunistic scheduling schemes	
	Less inter-cell interference because interference is now useful signal or reduced	
	SINR losses from running distributed network with handover margins to prevent ping-pong overhead can be reduced	
Limitations	Coherent transmission relies on very accurate multi-way channel estimation and feedback to the transmitter side	Own-cell resources are reduced to serve users in CoMP set which has impact in high-load scenario. This increases with higher CoMP set sizes
	Signal imbalance between cells in CoMP set should be below some threshold dB for significant benefit in downlink taking into account reduction in scheduling flexibility (does not apply for uplink)	In some heterogeneous network settings, eICIC may provide larger network gains than enabling JT CoMP
	Advanced UE reception techniques such as Interference Rejection Combining (IRC) partially exploit the same gain mechanisms as JT-CoMP	

depend on a tight path loss balance between the cells in the CoMP set; that is, less than 6–9 dB is needed for significant single-user benefit. In practice it seems that going beyond three cells in a CoMP set from a single user's perspective has limited benefit. In a 3GPP macro simulation environment with pathloss 6 dB restriction, the average number of users served in CoMP mode is approximately ~36% if 57 sectors are assumed in one cluster,[1] and ~11% CoMP UE corresponding to only three intra-site sectors in the CoMP cluster. Some characteristics of CoMP gains seen from single-user perspective and remaining users are listed in Table 13.4.

[1] Depends on the shape of the CoMP cluster.

13.4.3   Other Implementation Aspects

Beside the way how user data is distributed among cells, the schemes also differ substantially in various other requirements posed on the network, which we will address in the following.

13.4.3.1   Transport/Network Architecture

Different downlink CoMP schemes have different requirements on processing structure. JT benefits significantly from having centralized baseband processing, as shown in Figure 13.3, as two cells are supposed to transmit exactly the same data flow. This indicates that high transport capacity such as CPRI is required to carry the encoded and modulated data samples between the baseband pool and remote RF units. On the other hand, CS/CB can work in a distributed baseband structure, which means moderate requirements on transport capacity. DCS requirements are somewhere in between JT and CS/CB as on one hand these schemes require user data available at all involved notes, but on the other hand they don't require strictly synchronized transmission between cells.

Also, requirements will differ strongly depending on whether we are looking at intra-site or inter-site CoMP. Intra-site CoMP refers to the CoMP operation between different sectors of one base station. Intra-site CoMP does not set any additional requirements for the transport network nor for the synchronization. Inter-site CoMP sets high requirements, but gives also more benefit when CoMP operation can be extended also between separate base stations. The simulation studies in later sections show that intra-site CoMP can be an attractive option since it can provide most of the performance gains while keeping the network requirements low.

13.4.3.2   Base Station Synchronization

As in the cases of CS/CB and DCS only one eNodeB transmits to one user at the same time, these schemes do not pose other requirements on synchronization than an ordinary LTE system. However, a much more precise synchronization in time and frequency is needed in the case of JT, as only then a precise coherent overlap or destruction of signals at the terminal side can be achieved. The synchronization requirements for JT are approximately five times tighter than for CS/CB or DCS in order to achieve benefits from JT. We also note that the phase noise is one factor that should be minimized for best performance.

The requirements and characteristics of the three downlink CoMP schemes discussed in this section are summarized in Table 13.5.

13.5   Uplink CoMP Reception

As opposed to downlink, uplink CoMP reception is simpler from a 3GPP specification point of view. Uplink CoMP can be applied already for Release 8 UEs as uplink CoMP does not necessarily require any changes to the radio specifications. Naturally, the actual implementation of uplink CoMP requires changes on the network side. CoMP reception requires that the signals from several antennas in the network are routed to a central baseband receiver, where they can be combined and used in the detection process for every user. Joint reception techniques may be combined with other uplink enhancement techniques such as Interference Rejection Combining (IRC), adaptive antennas and multi-user detection schemes.

Table 13.5 Differences between joint processing and coordinated scheduling/beamforming

	Coordinated Scheduling and Beamforming (CS/CB)	Joint Processing (JP)	
		Dynamical Cell Selection (DCS)	Joint Transmission (JT)
Data availability	At one cell only	At all cells in CoMP set	At all cells in CoMP set
Data transmission	Always from serving cell	Coordinated transmission from single cell at a time	Coherent transmission from multiple cells
Transport	Only control requirements	Two times more user plane capacity required for CoMP users	Substantially high requirements for bandwidth and latency, see Table 13.2
Base station synchronization	0.05 ppm frequency and 3 μs timing accuracy	0.05 ppm frequency and 3 μs timing accuracy	Accurate synchronization required: 0.02 ppm frequency and 0.5 μs timing accuracy
Channel state information (CSI)	Lower requirements	Lower requirements	Accurate CSI required

As uplink CoMP reception can be done with no negative impact to system capacity in general, it is an attractive technique up to the point where it is too expensive to route the sampled signals via transport or to handle the additional computational complexity in the receiver. As such, the intra-site LTE CoMP is similar to softer handover in WCDMA. Intra-site CoMP sets no requirements for the transport. But uplink CoMP reception may also be done across a large set of cells when the baseband hotel model is applied, see the previous section.

There is a myriad of uplink CoMP techniques [2], each presenting a different compromise of performance gains, transport data rate and latency, as well as computational complexity. Some of the possible uplink CoMP techniques are listed in Figure 13.6.

In Figure 13.6, uplink CoMP techniques are categorized into scheduling and reception methods. In dynamic reception point selection, the most suitable reception point or cell for uplink can be selected based on even instantaneous channel state information. As more accurate cell selection is effectively achieved, cell edge data rates are improved. CoMP allows also for the selection of uplink reception points independently from downlink cell selection. This is especially beneficial in heterogeneous networks, where downlink transmit power is considerably smaller in small cells than in macrocells. On the other hand, the UE has the same transmit power in all cases which means that downlink coverage of a small cell can be smaller than its uplink coverage. In other words, a UE on the cell edge between macro- and small cell receives stronger downlink signals from a macrocell, but its uplink is received with better quality via the small cell. This is illustrated in Figure 13.7.

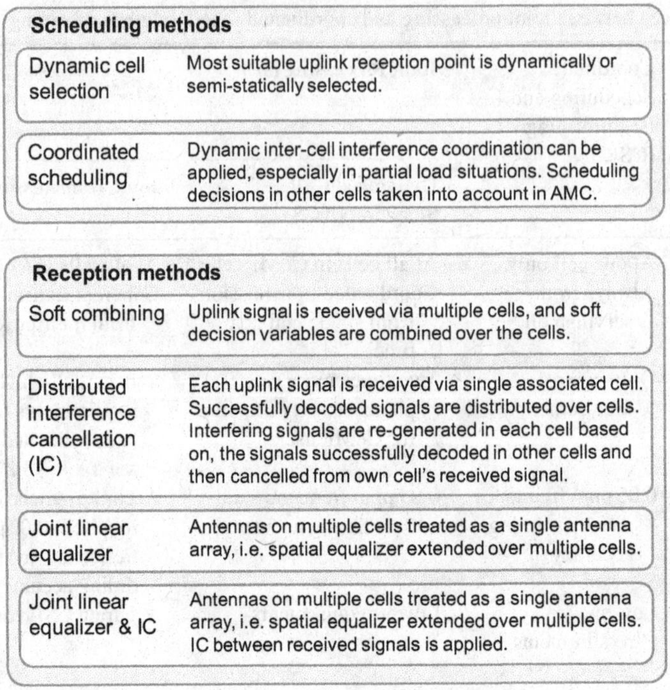

Scheduling methods	
Dynamic cell selection	Most suitable uplink reception point is dynamically or semi-statically selected.
Coordinated scheduling	Dynamic inter-cell interference coordination can be applied, especially in partial load situations. Scheduling decisions in other cells taken into account in AMC.

Reception methods	
Soft combining	Uplink signal is received via multiple cells, and soft decision variables are combined over the cells.
Distributed interference cancellation (IC)	Each uplink signal is received via single associated cell. Successfully decoded signals are distributed over cells. Interfering signals are re-generated in each cell based on, the signals successfully decoded in other cells and then cancelled from own cell's received signal.
Joint linear equalizer	Antennas on multiple cells treated as a single antenna array, i.e. spatial equalizer extended over multiple cells.
Joint linear equalizer & IC	Antennas on multiple cells treated as a single antenna array, i.e. spatial equalizer extended over multiple cells. IC between received signals is applied.

Increasing gains, transport data rates, computational complexity

Figure 13.6   Uplink CoMP techniques.

Scheduling that is coordinated over multiple cells allows for inter-cell interference-coordination on the time scale of individual scheduling decisions. This improves the efficiency of ICIC methods even further. Cell edge data rates can be improved by scheduling UEs in neighbouring cells so that severe inter-cell interference is avoided on uplink resources assigned for a cell edge UE. Inter-cell interference may also be partially predicted based on scheduling decisions made in neighbouring cells. This can be taken into account in the

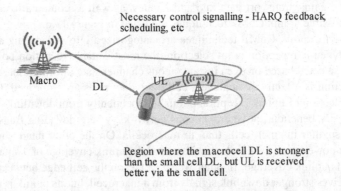

Necessary control signalling - HARQ feedback, scheduling, etc

Macro   DL   UL

Region where the macrocell DL is stronger than the small cell DL, but UL is received better via the small cell.

Figure 13.7   Independent uplink (UL) and downlink (DL) selection in a heterogeneous network.

selection of adaptive modulation and coding (AMC) schemes, hence, improving link adaptation accuracy.

CoMP scheduling methods can be applied together with conventional single cell reception methods. Hence, they do not necessarily require as large transport data rates as CoMP reception methods. They can be equally applied with multi-cell reception methods. For example, coordinated scheduling that avoids harsh inter-cell interference for cell edge UEs can be enhanced further with soft combining over the cells. This means that the signals from a certain cell edge UE are received by antennas from multiple sectors and soft decision variables are then jointly combined over the cells.

As a different variant of joint reception methods, it is possible to apply multi-cell joint processing but where UEs are still scheduled in each cell. In this case we allow users in different cells to utilize the same physical resources and we apply for example interference cancellation (IC) methods or spatial separation techniques. In this case, the scheduler algorithm will attempt to create good user pairs for efficient reception. There exist numerous possible reception methods. In principle, all antennas in the coordinated cells can be seen to form a single large antenna array over the cells. This means that all variants of MIMO receivers can be considered, ranging from linear multi-cell equalizers to sophisticated interference cancellation receivers, such as turbo equalizers with IC feedback based on turbo decoder output. Dimensions of such a single joint antenna array grow fast with increasing number of cooperative cells, and computational complexity of related receivers grows very fast beyond practical feasibility. Hence, we need to consider in practice sub-optimal receivers where signals from a certain UE are received via only a restricted sub-set of all available antennas.

To give an example on such suboptimal reception techniques, distributed interference cancellation is listed in Figure 13.6. In that, each uplink signal is received via antennas of a single cell. Signals are separately decoded in each cell at the first reception stage. After that, soft decoder output or hard bits of successfully decoded signals are distributed over cells. In those cells where decoding had failed, a second reception stage with inter-cell interference cancellation is entered. Interfering signals are re-generated based on decoded information from other cooperating cells and then cancelled from the own cell's received signals to enhance the detection of the desired signal. Interference re-generation requires that channels are also estimated for interfering signals based on reference signal information that is exchanged between cells.

In the case of centralized baseband hotels, the number of cells involved in a CoMP uplink may be rather large. Computational as well as system complexity for joint reception over all potential CoMP cells can easily become unfeasible even for sub-optimal joint receivers. The overall set of CoMP cells can be divided into a set of smaller CoMP cell clusters based on cell geometry. A more efficient but also computationally more complex joint reception method can be employed within CoMP clusters, while a computationally simpler uplink CoMP technique, such as coordinated scheduling, may be applied between CoMP clusters. Essentially, this means that different layers of inter-cell coordination are employed within a single centralized baseband hotel in a hierarchical manner. Hence, this is referred to as hierarchical CoMP techniques.

All CoMP reception techniques are based one way or another on additional signal information received on the neighbouring cells. This means that corresponding channels need also to be estimated on these neighbouring cells. That is a challenging task as the desired signal is typically received at a lower signal power on the neighbouring cells than the own signals in neighbouring cell. It has been shown that uplink CoMP reception is affected more by channel estimation than normal single cell reception [3]. Basically, it is possible to use CoMP reception with the default Demodulation Reference Signal (DM RS) configuration, where a sequence group is assigned separately for each cell and reference signals are randomized between cells. However, the use of inter-cell non-orthogonal DM RS significantly limits the CoMP reception gains. CoMP reception gains can be boosted by re-using the same sequence group over multiple cells in order to support inter-cell orthogonal DM RS. This allows for better multi-cell channel estimates, but the downside is the additional scheduler complexity, as the use of inter-cell orthogonal DM RS imposes some Physical Resource Block (PRB) allocation restrictions.

However, the standardization still progresses on improving uplink CoMP reception. Release 11 contains methods to improve CoMP performance on PUSCH and PUCCH. More flexible configuration of reference signals in UE-specific manner is introduced for both channels. For example, inter-cell orthogonal DM RS can improve CoMP performance and is supported already in Release 8. However, it involves scheduling limitations as PRB allocations need to be aligned for orthogonal DM RS. Such scheduling limitations can be alleviated by introducing UE-specific configuration of DM RS sequences and the related hopping pattern. Other methods that were considered during Release 11 include also power control enhancements.

13.6 Downlink CoMP Gains

The gain from CoMP on downlink performance is mainly in three major ways:

1. Inter-cell spatial domain coordination to suppress interference and/or improve the desired signal strength. This approach is sensitive to the channel state information accuracy therefore it poses tough requirements on the infrastructure.
2. Coordinate the radio resource management to optimize the resource assignment.
3. Mitigating the cell selection loss at the cell edge which is inherently there in distributed base station deployment with non-instantaneous cell selection mechanisms (due to handover margins and triggers).

The second way doesn't include precise spatial domain information; therefore it can provide moderate but robust gain and can work on ordinary infrastructure. CS/CB and DCS are mainly focusing on the second benefit while joint transmission is focusing on the first benefit. The third benefit can be significant at the cell edge (e.g. some implementations may operate with effective handover margins of 3–4 dB) but over time distributed solutions will also be optimized and will find ways to squeeze the effective handover margins. The CoMP gains mechanisms are shown in Table 13.6.

Figure 13.8 summarizes the observed CoMP gains over Release 8 2×2 SU-MIMO, where several downlink CoMP schemes were simulated in a typical homogeneous macrocell network layout with 500 m inter-site distance and with a full buffer traffic model. In the figure, intra-site CoMP means that the coordinated set is constrained to three co-sited sectors while inter-site refers to CoMP reception over different sites, each having three sectors. Cross-polarized antenna configurations are assumed for all cases. The two rightmost cases assume un-quantized CSI available at the eNodeB to exploit the maximum potential of

Table 13.6 Downlink CoMP gains with different CoMP options

	Coordinated Scheduling and Beamforming (CS/CB)	Dynamical Cell Selection (DCS)	Joint Transmission (JT)
Inter-cell spatial domain coordination	—	—	Yes
Coordinated radio resource management	Yes	Yes	Yes
Mitigation of handover margins	Yes	Yes	Yes

CoMP and all other cases assume quantized CSI (codebook) knowledge at the eNodeB side to emulate a realistic environment. The overhead assumption for DCS and Release 8 SU-MIMO are 2 Cell Specific Reference Signal (CRS) ports, for JT, CS/CB and Release 10 MU-MIMO is 2 CRS ports + 1/2 DMRS port.

The results show that DCS is the best downlink CoMP scheme under two-transmit antenna assumption. The radio resource can be utilized in more optimal way with DCS due to fast mobility management and fast resource coordination. The CoMP gain is still only about 10% for the cell edge users and no gain for the average users. JT is not able to provide too much signalling processing gain since the number of antenna dimension with two antennas is too restricted. The additional DMRS overhead of approximately 10% makes JT less attractive.

The use of four transmit antennas improves the downlink performance (especially coverage) – that is also without CoMP. With four transmit antennas, with more antenna dimension freedom, also the gain from JT can compensate the additional overhead and bring improved all gain mainly on cell edge perspective. The cell edge coverage gain of up to 40%

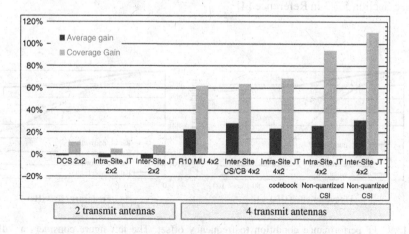

Figure 13.8 Downlink CoMP gains over Release 8 MIMO 2 × 2.

over Release 10 4 × 2 MU-MIMO is the upper bound gain from all CoMP schemes. As for the 2 × 2 case, the impact to average cell capacity is limited. However, when considering only quantized CSI available at eNodeB side, JT gain becomes close to 10%.

Beside imperfect channel knowledge, imperfect synchronization of eNodeBs in frequency is an inevitable impairment from hardware implementation. The frequency oscillator at each base station is subject to a certain extent of phase noise which results in random fluctuations in frequency. The requirement from RAN4 on inter-eNodeB frequency synchronization is 0.05 ppm in macro areas, which means the maximum frequency fluctuations of 200 Hz between two eNodeBs for a carrier frequency of 2 GHz. To make this number more illustrative, a 200 Hz frequency offset is similar to the Doppler frequency shift a UE moving at a speed of ~50 kmph is subject to, which is a big challenging to any closed loop operation.

Let us now assume a joint CoMP transmission among two cells to one terminal, where the two involved base stations have a frequency offset of Δf Hz. The impact of this offset on the transmission can be modelled by multiplying the signals originating from one of the two base stations by a factor of $e^{-j\Delta f t}$. Here, t is the time between the moment when the downlink channels are estimated, and when channel knowledge is finally applied at the transmitter side for joint multi-cell precoding. Hence, depending on the delay, the signals originating from one of the eNodeBs will experience a certain phase shift, which will impair a coherent overlap of signals at the receiver side. Such phase shift is difficult to be compensated from terminal side as it needs to know the exact frequency fluctuation value and also when the eNodeB will use this feedback. Neither of them can be easily known by UE side.

Figure 13.9 illustrates the JT performance difference depending on different frequency offsets. Here, four stream LTE codebook is used to quantize the CSI of two cells each with two antennas. Depending on whether rank 1 or rank 2 transmission and also whether wideband or narrowband CSI feedback is used, the performance degradation due to the frequency offset could be around 5–40%. In order to keep the performance degradation at low level, the frequency offset should be below 20–40 Hz which implies 5–10 times higher frequency synchronization requirement compared to the non-CoMP operation. We also note that rank 2 and wideband PMI are less sensitive to the frequency offset as both of them has less accurate spatial information included. Rank 2 codebook is much more coarse compared to that of rank1; see Section 5.3.3 in Reference [4].

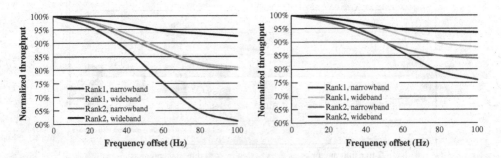

Figure 13.9 JT performance condition to frequency offset. The left figure considers a 3 dB SNR (signal power from 2 eNBs) and the right figure assumes a 13 dB SNR.

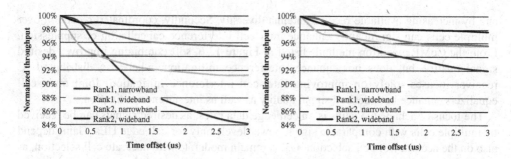

Figure 13.10 The degradation in link throughput as a function of timing error. The left figure considers a 3 dB SNR (signal power from two eNBs) and the right figure assumes a 13 dB SNR. Note that a 47 sample timing offset is equivalent to 3 μs.

Besides the frequency domain imperfection, time domain synchronization error is also a factor which cannot be ignored. For reference, the RAN4 requirement for inter-eNB timing offset is 3 μs in TDD while FDD does not need any time synchronization. The timing error plus the time delay difference from multiple transmission points to a UE will contribute to additional frequency selectivity in the case of joint transmission. Normally in small cells with site distance below 1 km, timing offset is the dominating factor. The specification of sub-band CQI/PMI has been proposed as a mechanism to counter this effect. Frequency selectivity can be compensated as far as the feedback granularity is finer than the coherence bandwidth. However, a practical system can only afford a certain level of feedback granularity which means larger overhead. In Figure 13.10 we illustrate the degradation of JT performance as a function of timing offset (10 MHz bandwidth, 47 samples correspond to 3 μs). As expected for timing offset of the order of a few samples, narrowband feedback is more robust to wideband feedback. So, the narrowband feedback can resist a certain degree of frequency selectivity. However, for timing offset beyond 5–10 samples, narrowband feedback also shows significant degradation in performance.

From this simulation, we can conclude that the timing synchronization should be kept within 0.3–0.5 μs. It's significantly higher than normally TDD requirement.

We can conclude that downlink CoMP is not a game-changer in terms of meeting the capacity growth demands.

- CoMP gain is up to 10–20% considering a realistic CSI knowledge. Intra site benefits constitute most of the total potential. Future macrosite solutions, like six sectors and more spectrum with carrier aggregation, offer great potential for operator that stay with distributed architecture.
- CRS based DCS can outperform JT in case of two transmitting antennas scenario considering the DMRS overhead.
- Additional synchronization requirements for are needed for JT.

13.7 Uplink CoMP Gains

Uplink CoMP reception techniques can improve uplink performance basically in two ways. Firstly, the desired signal quality can be improved by simply collecting more signal energy and by increasing available receive antenna diversity. Secondly, coordinated reception over multiple cells can significantly enhance inter-cell interference cancellation or suppression. From the CoMP reception methods listed in Figure 13.6, soft combining improves desired signal quality, but does not enhance interference mitigation, whereas a distributed IC receiver focuses solely on improving inter-cell interference mitigation. Joint multi-cell equalizers enhance both desired signal quality as well as interference mitigation.

The focus of achievable gains is on cell edge data rates, as desired signals can be received by multiple cells with comparable signal power levels only for cell edge UEs. Gains depend also on the accuracy of cell selection. Gains remain moderate with accurate cell selection, as received signal powers in neighbouring cells are below the received signal power level at the selected cell. Gains, of course, increase in case of less optimal cell selection. One particular case is seen with heterogeneous networks, where cell selection is a complex trade-off between downlink signal quality and cell splitting gains, as well as with uplink signal quality. As base station transmission power is considerably lower in small cells than in macrocells, small cell downlink coverage is significantly smaller than its uplink coverage. As UEs select cells based on downlink signal strength, uplink signal quality may be significantly improved for cell edge UEs simply by receiving it via the small cell, even though it is served in the downlink via the macrocell. This alone has resulted up to 40% improvement on cell edge data rates in simulations where four 4 pico cells are located within a macrocell and pico cell selection is biased 6 dB over macrocell selection. These CoMP reception gains depend heavily on the cell selection parameters; gains decrease if small cell selection is biased in the cell selection and vice versa.

Uplink CoMP enhancements on inter-cell interference mitigation become more important in interference limited environments. This is often the case in LTE due to the frequency reuse of one. There is also an important inter-play between interference mitigation and transmit power control. If the target of uplink CoMP reception is to improve cell edge data rates, power control is not significantly affected. Enhanced interference mitigation capabilities are used to clean up the signal received from cell edge UEs, while power control is used to keep balance between reasonable transmission powers and interference coming from (or generated towards) cells outside the CoMP reception area. Achieved CoMP gains focus on cell edge performance, although also cell capacity is increased. However, if the target is to boost cell capacity, the scenario is different. Transmit powers can be increased in general, and cell capacity is boosted by larger received signal powers as well as by enhanced interference mitigation. With single cell reception this would lead to significantly degraded cell edge performance due to increased interference levels. With CoMP reception, the enhanced interference mitigation capabilities are used to maintain cell edge data rates acceptable despite of considerably increasing interference originating from cells both inside and outside the CoMP reception area.

Uplink CoMP reception was simulated in homogeneous macrocell network layout with 500 m inter-site distance with full buffer traffic model. Observed CoMP gains for cell edge data rates are shown in Figure 13.11. Intra-site CoMP reception refers to CoMP reception over three sectors of single base station site, while inter-site CoMP reception refers to CoMP reception over three base station sites, each having three sectors. Fractional path loss compensation with α set to 0.8 was used in power control both with and without CoMP. A multi-cell joint equalizer with an IC stage was used as a CoMP receiver, and it was compared against a two receive antenna single cell receiver with interference rejection combining

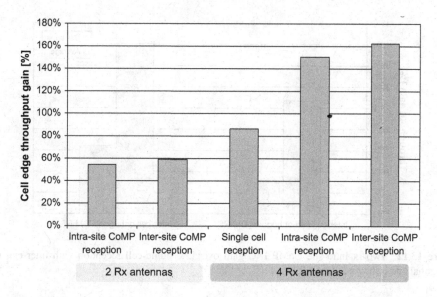

Figure 13.11 Uplink CoMP IRC-SIC reception gains on cell edge data rates compared to 2rx single cell reception.

(IRC) without inter-cell orthogonal DM RS. Results show that cell edge data rates can be considerably boosted already with intra-site CoMP reception, that is, with sophisticated signal processing within single site. It should also be noted that cell edge performance is boosted more by increasing the number of receive antennas from two to four antennas per cell. CoMP reception provides similar relative cell edge gains over single cell IRC receivers both in two and four receive antenna cases.

When considering reception methods listed in Figure 13.6, soft combining provided only modest CoMP gains as its gain mechanism is limited to improved desired signal collection but it cannot improve inter-cell interference suppression further. On other hand, distributed interference cancellation as well as multi-cell joint linear equalizer provided reasonable CoMP gains. Both reception methods provided rather comparable gains and delivered most of the gains achievable with multi-cell joint equalizer with IC stage.

Inter-cell orthogonal DM RS are assumed in Figure 13.11. The impact of inter-cell orthogonal DM RS is illustrated in Figure 13.12 for intra-site CoMP reception with two receive antennas and IRC CoMP receiver. Compared to Figure 13.12, the receiver is slightly different as no IC stage is included, and this degrades the performance gain over the reference by about 10%. Cell-specific sequence groups with appropriate inter-cell randomization were used as inter-cell non-orthogonal DM RS configuration. In the case of inter-cell orthogonal DM RS configuration, a single sequence group was used within CoMP coordination area. Cyclic shifts as well as orthogonal cover codes were used to create orthogonal reference signals. To support orthogonal reference signals, PRB allocations were aligned over the cells in both cases. It can be noted that inter-cell orthogonal DM RS improves considerably CoMP reception gains by enhancing multi-cell channel estimation accuracy.

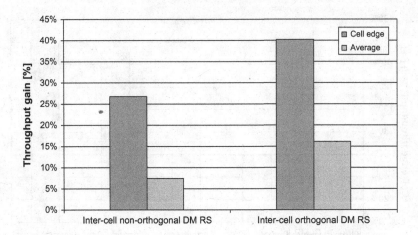

Figure 13.12 Two rx intra-site CoMP IRC gain over 2rx single-cell reception with inter-cell non-orthogonal and orthogonal DM RS.

13.8   CoMP Field Trials

A variety of downlink and uplink CoMP concepts have already been successfully tested in field trials. In the research projects EASY-C and Artist4G, partially funded by the German Ministry for Education and Research (BMBF) and the European Union, respectively, a large cellular test system for LTE-Advanced has been setup in downtown Dresden, Germany.

One of the downlink CoMP schemes tested was a joint transmission of up to three cells with two antennas each to three terminals with two receive antennas each. Here, it could be shown that if the terminals are fairly static and located close to the cell edge between the three cells, joint transmission can strongly mitigate the effect of inter-cell interference and consequently boost the data rates of all three terminals. The successful trials also showed that a sufficient synchronization of base stations in frequency domain can be obtained, and that it is in principle possible to feed channel information back from the terminals to the infrastructure in sufficient quality. In fact, the downlink CoMP gains measured in field trials were larger than stated in this chapter, but this may be explained by the fact that in the down-link trials, only the subset of base stations involved in a joint transmission was turned on. Hence, there was no background interference from other base stations, as is typically assumed in system level simulations.

To evaluate uplink CoMP concepts in practice, two terminals were driven along a trajectory encompassing 16 base stations, and joint multi-cell detection was compared against distributed interference cancelation and a non-cooperative baseline assuming interference rejection combining. The results obtained are very much in line with the simulation results discussed earlier. The results of both downlink and uplink field trials were documented in [5].

Field trial results cannot replace system level simulations, as one cannot test the behaviour of a fully loaded commercial system, but they can provide the proof-of-concept that certain technologies principally work. In the case of CoMP, this has been done successfully, and this has also led to the fact that CoMP was included into Release 11.

13.9 Summary

Practical LTE data rates are limited by inter-cell interference rather than by thermal noise. In particular, current systems bear the problem that cell edge data rates are often just a fraction of the theoretical peak rate. Coordinated Multipoint (CoMP) is an option to manage inter-cell interference and boost cell edge data rates, and consequently provide a more homogeneous quality of service from cell centre to cell edge.

In the downlink, a joint transmission from multiple cells to one or multiple terminals has the highest gain potential, but requires very low latency backhaul while coordinated scheduling and beamforming sets lower requirements. Intra-site downlink CoMP between multiple sectors is the most straightforward option without any further backhaul requirements and achieves the majority of the traditional CoMP gains. The downlink cell edge CoMP gains with four antenna base station transmission and including channel state imperfections are up to 10–20%. The downlink CoMP gains with two antenna transmissions are clearly lower than with four antennas. Further, the downlink CoMP is not a game-changer in terms of meeting the capacity growth demands since the average gains tend to be only modest. The simulations show that intra site CoMP benefits constitute most of the total CoMP gains. Therefore, intra-site CoMP could be an attractive option to enjoy CoMP gains without major changes to the network architecture. Further sectorization and additional RF heads can make the single site configuration larger and make intra-site CoMP more interesting.

In the uplink, a variety of CoMP schemes exist, from coordinated scheduling over distributed interference cancellation to joint multi-cell detection. While the latter schemes provide significantly larger gains, the price is that a large backhaul capacity is required. Simulations have shown that uplink joint detection can provide cell edge data rate increases of 40–50%. The gain is even higher for the heterogeneous networks if the uplink signal can be received both via a small cell and a large macrocell. Simulations show that intra-site CoMP provides the majority of the total CoMP gains allowing also significant benefit of CoMP technology in a traditional and distributed network deployment.

Uplink CoMP can be implemented with Release 8 terminals while downlink CoMP requires new terminals. Downlink CoMP and further uplink CoMP enhancements is included into Release 11 specifications, with some enhancements foreseen to continue in Release 12.

References

1. Tse, D. and Viswanath, P. (2011) *Fundamentals of Wireless Communications*, Cambridge University Press.
2. Marsch, P. and Fettweis, G.P. (2011) Uplink CoMP under a Constrained Backhaul and Imperfect Channel Knowledge, *IEEE Transactions on Wireless Communications*, **10** (6), 1730–1742.
3. 3GPP R1-112393 (2011) Channel estimation modeling for uplink in system level simulation. Nokia Siemens Networks, Nokia, 3GPP RAN1#66 meeting, August 22–26, 2011, Athens, Greece.
4. 3GPP TS 36.211 (September 2011) Physical Channels and Modulation, V.10.3.0.
5. Marsch, P. and Fettweis, G.P. (eds) (2011) *Coordinated Multipoint in Mobile Communications*, Cambridge University Press.

14

HSPA Evolution

Harri Holma and Karri Ranta-aho

14.1 Introduction

High Speed Packet Access (HSPA) made data services go wireless in a massive way. The mobile networks mainly carried voice traffic before HSPA was launched while now HSPA traffic is more than ten times voice traffic in many markets. The fast growth in data usage and the new smartphone applications set higher requirements also for HSPA radio. A large number of smartphones with HSPA radio are expected to hit the market pushing the need for further HSPA evolution. This chapter presents the main features and the benefits in HSPA evolution. The features are divided into the following sections: multicarrier, multiantenna and multicell evolution, packet data enhancements, voice evolution, advanced receivers, flat architecture and LTE interworking. WCDMA and HSPA evolution is presented in more detail in [1].

14.2 Multicarrier Evolution

The most straightforward way of increasing the data rates is to use more bandwidth. Dual cell (dual carrier) HSDPA (DC-HSDPA) in Release 8 uses two adjacent carriers in downlink to push the data rates up twofold. DC-HSDPA networks were widely launched during 2011. The bandwidth can be further increased to 20 MHz in Release 10 by using four carriers and even to 40 MHz in Release 11 with eight carriers. The uplink dual cell was included in Release 9 specifications. The multicarrier evolution is illustrated in Figure 14.1.

Operators have deployed multiple carriers in their networks almost from the very beginning of WCDMA for capacity reasons. Harnessing the already-deployed multicarrier infrastructure to delivering data to one user is a cost-efficient way for increasing the system peak data rates. The spectrum utilization is also significantly improved because it is possible to efficiently balance the loading between carriers and reduce the probability that some carriers don't see any loading while some other carriers are fully utilized due to a momentary high-data rate download. Multicarrier links also enable frequency domain scheduling,

LTE-Advanced: 3GPP Solution for IMT-Advanced, First Edition. Edited by Harri Holma and Antti Toskala.
© 2012 John Wiley & Sons, Ltd. Published 2012 by John Wiley & Sons, Ltd.

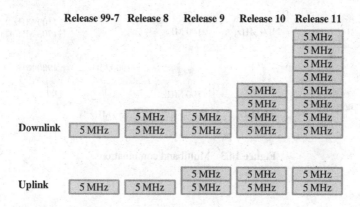

Figure 14.1 Multicarrier evolution.

providing system gains even if all carriers in the system are fully utilized. Notably the system benefits of improved spectrum utilization, more dynamic load balancing and frequency domain scheduling gains are available even if only a fraction of the devices in the system are multicarrier capable.

Figure 14.2 illustrates the individual user data rate distribution of a single, four-carrier and eight-carrier aggregation, when the users are experiencing a 1-megabit file arriving with a random probability, the mean file inter-arrival time being one file every 5 s. When a traffic

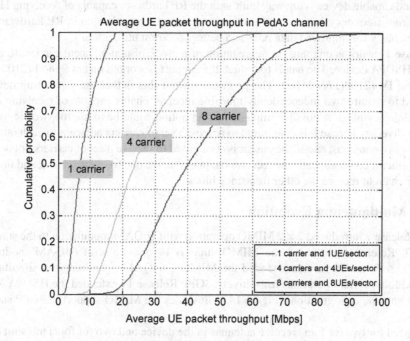

Figure 14.2 Multicarrier performance simulations.

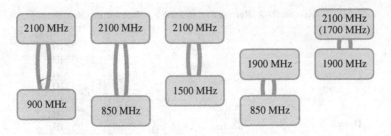

Figure 14.3 Multiband combinations.

burst starts, there is a high likelihood that there are multiple carriers that are currently under-utilized in the system. A single carrier user is nevertheless only able to receive data over one carrier leaving the other carriers unused. A multicarrier user is however able to benefit from the free capacity on multiple carriers and thus experiences a higher data throughput. The median data rate with single carrier is 7 Mbps, with four-carrier 24 Mbps and with eight-carrier 40 Mbps. In short, the four-carrier solution in this case provides 3.4 times higher data rate and the eight-carrier 5.7 times higher data rate than the single carrier solution.

Release 9 also brought the possibility to allocate the two carriers from two different frequency bands to a DC-HSDPA device, and this was further extended in Release 10 so that the four carriers can be split between two frequency bands. Figure 14.3 illustrates the standard supported band combinations that can be aggregated to a multicarrier capable device. As a multiband capable device is anyway built with the RF hardware capable of receiving HSDPA on different frequency bands, it is attractive to aim at using that same RF hardware also simultaneously for improved data rates and better spectrum utilization.

Release 11 work is ongoing for aggregation of a block of non-adjacent carriers to a four-carrier HSDPA device. The bands for which the support is worked on are Band I (2100 MHz) and Band IV (2100/1700 MHz). The practical side of this is that the supporting device is expected to require two independently tuneable receiver chains capable of receiving on the same frequency band. A possible implementation option could be to use the two receivers as receiver diversity in case it is configured to receive a set of carriers adjacent to each other and it houses two antennas, and in the case it is configured to receive a set of carriers not adjacent to each other it could allocate one receiver chain to receive one frequency block and the other receiver chain to receive the other frequency block.

14.3   Multiantenna Evolution

3GPP Release 7 introduced 2×2 MIMO operation with 16QAM modulation to the standards in 2007, Release 8 extended the MIMO support to operate with 64QAM modulation and Release 9 and 10 further extended the MIMO + 64QAM operation with simultaneous two- and four-carrier operations respectively. 3GPP Release 11 extended the HSDPA MIMO to 4×4 antenna configurations. Figure 14.4 illustrates the MIMO support in 3GPP standard releases.

The need for two (or four) receive antennas in the device and two (or four) transmit antennas in the base station have somewhat slowed down the enthusiasm for practical MIMO

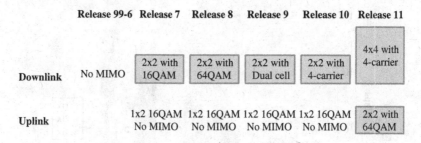

	Release 99-6	Release 7	Release 8	Release 9	Release 10	Release 11
Downlink	No MIMO	2x2 with 16QAM	2x2 with 64QAM	2x2 with Dual cell	2x2 with 4-carrier	4x4 with 4-carrier
Uplink		1x2 16QAM No MIMO	1x2 16QAM No MIMO	1x2 16QAM No MIMO	1x2 16QAM No MIMO	2x2 with 64QAM

Figure 14.4 Multi-antenna evolution.

deployments. Furthermore the additional overhead introduced in the MIMO enabled cell due to the second antenna pilot transmission slightly degrades the performance of the non-MIMO devices operating in the cell.

Figure 14.5 illustrates the virtual antenna mapping integrating the two transmit antennas of 2×2 MIMO with the single-transmit-antenna signals understandable to the non-MIMO devices. The *Pilot 2* and *Data 2* represent the MIMO signals only understood by the MIMO receivers. Both pilot signals need to be present for a MIMO UE to be able to estimate the channel quality and provide MIMO Channel Quality Indicator (CQI) reports in the uplink. Both data signals are only present when a MIMO UE is scheduled with dual-stream transmission. The virtual antenna mapper is used to split the transmit powers of the two virtual antennas between the two physical radio transmitters so that the cell's full transmit power capacity is available also for the non-MIMO transmissions.

Figure 14.6 shows the average cell throughputs with different antenna configurations. It can be seen that adding more receiver antennas is more beneficial than adding transmit antennas. Doubling the number of receiver antennas increases capacity by an average 50–60%. Doubling the number of transmit antennas increases capacity by 15–25%. The

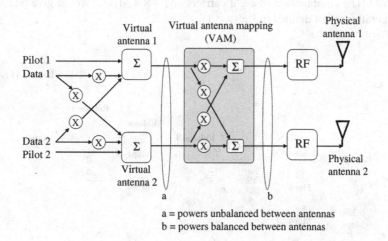

a = powers unbalanced between antennas
b = powers balanced between antennas

Figure 14.5 Virtual antenna mapping and 2×2 MIMO in the base station.

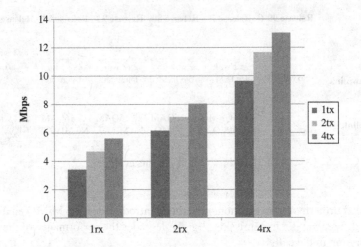

Figure 14.6   Average cell throughputs with different numbers of receiver and transmit antennas.

transmit antennas use feedback from UE to control the dual stream and transmit diversity usage. The feedback quantization and delay make the optimal transmission more challenging than with the receiver antenna diversity. The highest spectral efficiency is provided by 4×4 MIMO case with cell throughput of 13 Mbps in 5 MHz corresponding to the spectral efficiency of 2.6 bps/Hz/cell.

Figure 14.7 shows the HSPA peak data rate evolution through the 3GPP releases. The improvements in the peak data rates have been achieved by aggregating more carriers, using more transmit and receive antennas and transmitting more bits per symbol with higher order modulations.

The peak data rate in downlink is extended in Release 11 to 336 Mbps. That peak data rate can be achieved either by using eight carriers and 2×2 MIMO, or by using 4 carriers and 4×4 MIMO. The combination of eight carriers and 4×4 MIMO would give 672 Mbps but that configuration is not defined in Release 11.

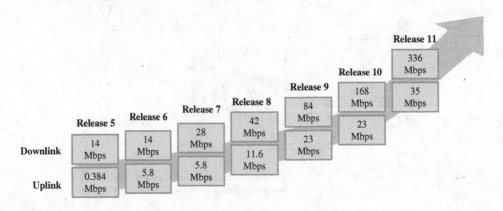

Figure 14.7   Peak data rate evolution.

3GPP Release 11 introduced multiantenna transmission also in the HSPA uplink. Both uplink beamforming (also referred to as single stream MIMO) for improved uplink coverage, and 2×2 MIMO are supported. 2×2 MIMO together with 64QAM modulation can boost the uplink data rate to 34.5 Mbps on a single 5 MHz carrier.

14.4 Multiflow Transmission

WCDMA Release 99 uses soft handover to improve the quality of the connection at the cell edge and to minimize the inter-cell interference. Soft handover is used also with HSUPA to control the uplink interference. But HSDPA was designed with single cell transmission without any soft handover. The single cell transmission makes the scheduling and fast retransmission control simple but the challenge is that cell edge users experience high inter-cell interference which limits the data rates. One approach to enhance the cell edge data rates is to transmit the data stream from two cells to one UE. This approach is called multiflow HSDPA transmission. The multiflow concept is illustrated in Figure 14.8. Multiflow can be upgraded to the network without any changes to the architecture since HSPA networks already include RNC that can act as the control point for the multiflow transmission. Multiflow is part of 3GPP Release 11.

Release 11 allows using multiflow together with dual cell HSDPA: the data will be sent from two separate sectors/base stations and up to four cells (up to two carriers) towards one UE. The multiflow can be done from multiple cells of the same site (intra-site multiflow) or from multiple sites (inter-site multiflow). In the intra-site case the data is split at NodeB in MAC layer and NodeB can run joint scheduling between the two cells in the same way as with dual cell HSDPA. In the inter-site case the data is split at RNC in the RLC layer and the scheduling is done independently.

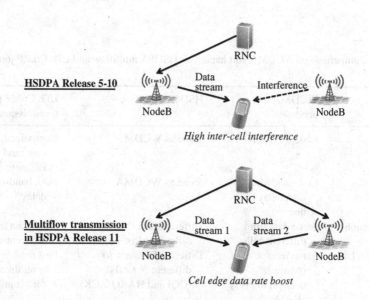

Figure 14.8   Multiflow transmission concept.

The scheduling can utilize prioritization so that normal priority is applied for the serving cell transmission while lower priority is used for the secondary cell. The target is that multiflow will not degrade the data rate of non-multiflow UEs but can still utilize unused resources in the adjacent cells.

HSDPA multiflow has some similarities with WCDMA soft handover and LTE Coordinated Multipoint Transmission (CoMP) with joint transmission. HSDPA multiflow can utilize the same architecture and transport as WCDMA while joint transmission CoMP uses centralized baseband and low delay transport. HSDPA multiflow splits the data transmission and different data is transmitted from different cells while in WCDMA soft handover and in LTE CoMP the same data is transmitted from both cells. The HSDPA multiflow exploits the uplink soft handover in its feedback; it transmits cell-specific CQIs and retransmission acknowledgements, and the base stations participating in soft handover for the data channels receive also all the multiflow feedback as well, and just ignore the parts that were meant to the other base station. The main differences between the multi-cell concepts are summarized in Table 14.1.

The cell edge data rate is improved with multiflow because of two factors:

a. More power is available per user when two cells transmit the signal towards UE.
b. Inter-cell interference can be managed more efficiently when UE is equipped with receiver that can reduce inter-cell interference (Type 3i).

Figure 14.9 shows the user data rate distribution in system simulations with two UEs per cell both with and without multiflow feature. The cell edge data rate (5% percentile) is improved by 33% from 3 to 4 Mbps. The cell edge improvements are important because the low data rates at the cell edge are typically the limiting factor from the end user point of view. The median data rate is also improved by 10% with the multiflow concept.

Table 14.1 Comparison of WCDMA soft handover, HSDPA multiflow and LTE CoMP joint transmission

	WCDMA soft handover	HSDPA multiflow	LTE CoMP joint transmission
Architecture	RNC + NodeB	Same as WCDMA	Centralized baseband + RF units
Transport	Low bandwidth high delay is fine	Same as WCDMA	High bandwidth low delay
Data transmission	Same data from different cells	Different data from different cells	Same data from different cells
Feedback from UE	Same feedback (power control) to every NodeB	Different feedback to different NodeBs (CQI and HARQ-ACK)	Fast feedback to the centralized baseband for beamforming

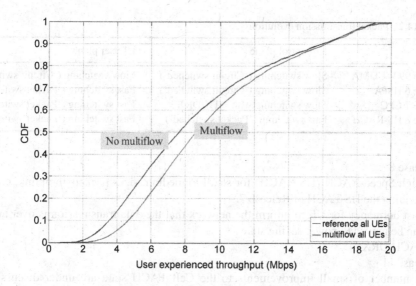

Figure 14.9 Multiflow transmission benefit.

14.5 Small Packet Efficiency

Much of the discussion has been allocated for improving the peak data rates and practical user data rates. Those data rate capabilities are important for large file transfer but further system features are required to support efficiently the transmission of small packets. HSPA in Releases 5 and 6 improved packet efficiency considerably by bringing fast scheduling and resource allocation to the base station. There are still some limitations in HSPA – mainly the continuous transmission of physical control channel DPCCH. Let's consider small packet sizes of 0.5–10 kB which are typical with smartphone applications. Those packet sizes are too large for RACH which can carry maximum a few hundred bytes by using multiple RACH messages. Therefore, HSPA channel need to be allocated often for smartphone traffic. The transmission time of 1 kB with 1 Mbps HSPA data rate is less than 10 ms. The allocation of HSPA channel takes several 100 ms, the inactivity release timer is a few seconds but the actual transmission time is just a few milliseconds, which is highly inefficient. The problem here is that the interference caused by the continuous DPCCH is relatively high, especially in uplink. The network resource consumption is high when the channel is occupied for a long time compared to the actual usage, and the UE power consumption increases when running DPCCH. The delay in setting up the channel impacts end user performance and the relative signalling overhead is high compared to the data volume. Several solutions were included into HSPA Releases 7 to 11 to improve the packet efficiency:

- Release 7
    - Continuous Packet Connectivity (CPC) which brings Discontinuous Transmission and Reception (DTX/DRX) for DPCCH in Cell_DCH state
    - High Speed FACH (HS-FACH) for small to medium sized packets in downlink without needing to move the UE to Cell_DCH state.

Table 14.2 Packet transmission evolution

	L1 control plane	L1 user plane
Release 99 WCDMA	Slow switching ('Circuit switched')	Slow switching ('Circuit switched')
Release 6 HSPA	Slow switching ('Circuit switched')	Fast switching ('Packet switched')
Release 7 CPC	Slow switching with DTX/DRX	Fast switching ('Packet switched')
Release 8 HS-RACH	Fast switching ('Packet switched')	Fast switching ('Packet switched')

- Release 8
  - High Speed RACH (HS-RACH) for small to medium sizes packets in uplink, complementing the HS-FACH of Release 7.
  - Fast dormancy for UE to inform the network that the data transmission is over and UE can be moved to power saving state
  - FACH DRX
- Release 11
  - A number of small improvements to the Cell_FACH state are under discussion in Release 11. The main ones include Node B initiated uplink setup in Cell_FACH for reduced uplink data latency and earlier availability of CQI and HARQ-ACK feedback for HS-FACH, enhanced DRX operation in Cell_FACH, and optimized Cell_FACH to Cell_PCH transition, each aimed at benefiting small packet delivery, or related signalling load and battery consumption.

The evolution of packet transmission is summarized in Table 14.2 and illustrated in Figure 14.10.

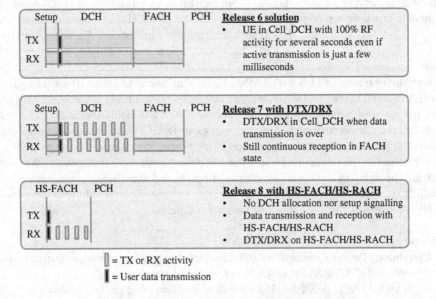

Figure 14.10 Evolution of small packet transmission solutions.

14.6 Voice Evolution

The good quality voice service is the key part of smartphone applications. HSPA evolution enables a number of improvements to voice services in the areas of voice quality, capacity, power consumption and LTE interworking.

14.6.1 Adaptive Multirate Wideband (AMR-WB) Voice Codec

AMR-WB was defined in 3GPP Release 4 for a circuit switched voice. AMR-WB enhances the voice quality by increasing the voice sampling rate from 8 to 16 kHz which makes the audio bandwidth larger. The typical radio data rate is still similar to AMR Narrowband (AMR-NB), so the radio capacity for AMR-WB is similar to AMR-NB. AMR-WB has been activated in tens of networks and many new UEs support AMR-WB.

14.6.2 Voice Over IP (VoIP)

3GPP Release 7 enhancements for HSPA make it possible to run good quality VoIP service with high capacity. HSPA radio brings the benefit that DTX/DRX features can be used also for voice which improves the UE talk time and reduces network interference leading to higher capacity especially in uplink. VoIP may also make it simpler to introduce new packet services.

14.6.3 CS Voice Over HSPA (CSoHSPA)

3GPP Release 8 enables to run CS voice over HSPA radio. This solution is similar to Voice over IP (VoIP) from the radio perspective and similar to CS voice from the core network perspective.

14.6.4 Single Radio Voice Call Continuity (SR-VCC)

SR-VCC enables handover between CS voice and VoIP. SR-VCC from VoIP to CS is defined in Release 8 and from CS to VoIP in Release 11. SR-VCC is needed especially in the early phase of LTE deployment when UE with VoIP call runs out of LTE coverage and needs to make handover from LTE VoIP to 3G CS voice.

14.7 Advanced Receivers

The practical data rates are limited by the interference. The data rates can be improved by advanced receivers that are able to manage interference more efficiently. The advanced receivers can be applied both in UE for the downlink and in NodeB for the uplink.

14.7.1 Advanced UE Receivers

Advanced UE receivers are able to tolerate both intra-cell interference and inter-cell interference. The intra-cell interference cancellation is called equalizer and is already included in most HSDPA receivers. 3GPP has defined performance requirements for different type of advanced receivers:

• Type 1 with receiver diversity;
• Type 2 with equalizer;

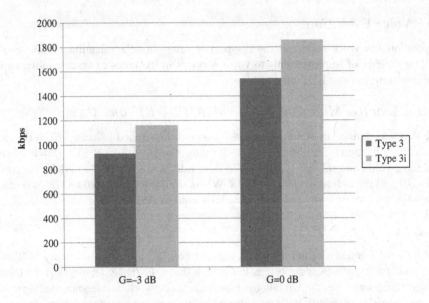

Figure 14.11   Throughput gain of the 3i type receiver [2].

• Type 3 with receive diversity and equalizer;
• Type 3i with receiver diversity, equalizer and inter-cell interference cancellation.

The receiver algorithms are not defined by 3GPP but are UE vendors specific and can also be better than 3GPP requirements. Simulation results for Type 3 and 3i are shown in Figure 14.11 [2]. The throughput gains with 3i compared to 3 are 20–25% for the cell edge conditions with G-factor of −3 dB and 0 dB. The live network measurements have illustrated even higher gains in practice. The advanced UE receives are nice from the operator point of view because there are no changes required to the network.

14.7.2   Advanced NodeB Receivers

Advanced NodeB receivers can remove intra-cell interference and own signal multipath interference. The inter-cell interference in uplink is so weak that it will not be cancelled. There are no 3GPP requirements for advanced NodeB receivers. One attractive solution is Turbo coded Parallel Interference Cancellation (PIC). The idea is to estimate the physical channel data after channel Turbo decoding the data. The data is re-generated by encoding the decoded data again and then removed from received signal to improve the quality of the reception for the other users. The simulations show up to 60% gain in uplink capacity with Turbo PIC. The receiver architecture is shown in Figure 14.12. The interference cancellation is important especially for high data rate HSUPA connection that would otherwise cause high interference to other simultaneous users.

The uplink equalizer is relevant for the high data rate HSUPA connection in multipath channels. The multipath interference impacts the quality of the reception and can be improved by using the equalizer receiver.

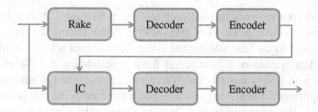

Figure 14.12 Receiver architecture for coded parallel interference cancellation (IC = Interference Cancellation).

14.8 Flat Architecture

3GPP Release 6 network architecture has four network elements in the user and control plane: base station (NodeB), RNC (Radio Network Controller), SGSN (Serving GPRS Support Node) and GGSN (Gateway GPRS Support Node). The architecture in Release 8 LTE has only two network elements: base station in the radio network and Access Gateway (a-GW) in the core network. The a-GW consists of control plane MME (Mobility management entity) and user plane SAE GW (System Architecture Evolution Gateway). The flat network architecture reduces the network latency and thus improves the overall end user performance. The flat model also improves both user and control plane efficiency. The flat architecture is considered beneficial also for HSPA and it is specified in Release 7. The HSPA flat architecture in Release 7 and LTE flat architecture in Release 8 are exactly the same: NodeB responsible for the mobility management, ciphering, all retransmissions and header compression both in HSPA and in LTE. The architecture evolution in HSPA is designed to be backwards compatible: existing terminals can operate with the new architecture and the radio and core network functional split is not changed. The architecture evolution is illustrated in Figure 14.13.

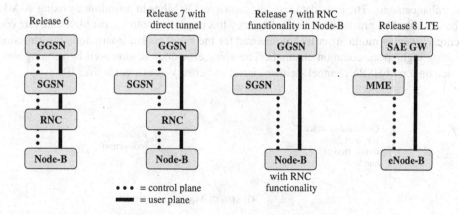

Figure 14.13 Evolution towards flat architecture.

Also the packet core network has flat architecture in Release 7. It is called direct tunnel solution and allows the user plane to by-pass SGSN. When having the flat architecture with all RNC functionality in the base station and using direct tunnel solution, only two nodes are needed for user data operation. This achieves flexible scalability and allows introducing the higher data rates with HSPA evolution with minimum impacts to the other nodes in the network. This is important for achieving low cost per bit and enabling competitive flat rate data charging offerings. As the gateway in LTE is having similar functionality as GGSN, it is foreseen to enable deployments of LTE and HSPA where both connect directly to the same core network element for user plane data handling directly from the base station.

14.9 LTE Interworking

The introduction of LTE on top of HSPA network requires interworking features in both radio networks. The interworking includes idle mode, voice handovers and data handovers. In the first phase of LTE deployment LTE capable UE makes reselection from HSPA to LTE. The reselection is allowed only in idle and in PCH states but not in FACH states. Therefore, it is important to set the HSPA parameters so that UE can get quickly to PCH state in order to enable smooth access to LTE network. In the later phase also redirection and packet handover will be supported which allows moving UE from HSPA to LTE also from Cell_DCH state. If UE runs out of LTE network, the LTE network can move it to HSPA network by redirection or handover. The shortest connection break is obtained with packet handover. The first phase interworking is illustrated in Figure 14.14.

The early phase voice service for LTE smartphones uses CS fallback handovers where UE is moved from LTE to 3G during the call setup phase. The network switching can utilize redirection or handover. When the voice call is over, UE returns back to LTE by reselection or by redirection.

14.10 Summary

HSPA evolution in Releases 7, 8, 9, 10 and 11 has introduced a large number of features for improving user data rates, spectral efficiency, voice support and smartphone support by using wider bandwidths, multiantenna solutions, multicell transmission and a number of packet data enhancements. The peak data rate has grown to 336 Mbps in downlink by using 40 MHz of bandwidth. The practical network efficiency has improved both for the high data rate connections by using multicarrier techniques and for the small packet smartphone transmissions by using high speed common channels. The voice efficiency is improved by running voice service on top of HSPA channel – either circuit switched voice or voice over IP.

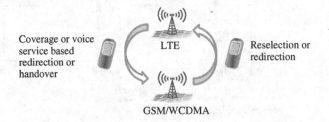

Figure 14.14 Early phase interworking between LTE and HSPA.

Table 14.3 Summary of performance benefits from HSPA evolution

	Peak data rate	Cell capacity and network utilization	Cell edge data rate	Smartphone experience
Multicarrier	++	++	++	+
Multiantenna	++	+ (tx antennas) ++ (rx antennas)	+ (tx antennas) ++ (rx antennas)	+
Multiflow	−	+	++	+
Small packet efficiency	−	+	−	++
Voice evolution	−	+	−	+
Advanced receivers	−	+	++	+

Table 14.3 summarizes the main HSPA evolution areas and the corresponding benefits for the different use cases. Multicarrier and multiantenna solutions are pushing the peak data rates. All these features improve the cell capacity and network utilization – especially multicarrier solution for the load balancing and more receiver antennas for improved link performance. Those features help also for the cell edge data rates but also multiflow and advanced UE receivers help for the cell edge users in downlink. The smartphone experience includes voice quality, low latency for smartphone applications and power consumption. HSPA evolution is ready to serve an increasing number of customers with an enhanced end user performance.

References

1. Holma, H. and Toskala, A. (2010) *WCDMA for UMTS – HSPA evolution and LTE*, 5th edn, John Wiley & Sons, Ltd, Chichester.
2. 3GPP TR 25.963 (2007) Technical Report, 3rd Generation Partnership Project; Feasibility study on interference cancellation for UTRA FDD, User Equipment (UE).

Index